The Same and Not the Same

THE SAME AND NOT THE SAME

Roald Hoffmann

Columbia University Press  NEW YORK

Columbia University Press
New York   Chichester, West Sussex
Copyright © 1995 Columbia University Press
All rights reserved

Library of Congress Cataloging-in-Publication Data
Hoffmann, Roald.
    The same and not the same / Roald Hoffmann.
      p.   cm. — (George B. Pegram lecture series)
    ISBN 0–231–10138–4.
    ISBN 0-231-10139-2 (pbk)
    1. Chemistry—Popular works. I. Title. II. Series.
QD37.H612   1995
540—dc20   94–39467

    Designer: Teresa Bonner
        Text: New Baskerville
Compositor: Graphic Composition
     Printer: Oceanic Graphic Printing
     Binder: Oceanic Graphic Printing

Printed in Hong Kong

c 10 9 8 7 6 5 4 3 2
p 10 9 8 7 6 5 4 3

The George B. Pegram Lecture Series

To provide an opportunity for distinguished scholars to examine the interaction between science and other aspects of our culture and society, the Trustees of Associated Universities, Inc., established the George B. Pegram Lecture Series at Brookhaven National Laboratory.

The lectures are named in honor of George Braxton Pegram, who contributed so much to our country in general, and to Brookhaven National Laboratory in particular. Except for a few years abroad, George Pegram's entire professional career was spent at Columbia University, where he was a Professor of Physics, Dean, and Vice President. In 1946, he headed the Initiatory University Group, which proposed that a regional center for research in the nuclear sciences be established in the New York area. Thus, he played a key role in the founding of Brookhaven and became one of the incorporating trustees of Associated Universities, Inc., remaining an active trustee for ten years.

George Pegram devoted his life to physics, teaching, and the conviction that the results of science can be made to serve the needs and hopes of humankind.

George B. Pegram Lectures

1949 Lee Alvin DuBridge	1972 Colin Low
1960 René Jules Dubos	1975 Jean Mayer
1961 Charles Alfred Coulson	1979 Sir Peter Medawar
1962 Derek J. deSolla Price	1985 David Baltimore
1963 J. Robert Oppenheimer	1988 Robert C. Gallo
1964 Barbara Ward	1989 Sir Denys Wilkinson,
1965 Richard Hofstadter	Michael S. Brown
1966 Louis S. B. Leakey	1990 Roald Hoffmann
1968 André Maurois	1992 Maurice Goldhaber
1969 Roger Revelle	1993 James D. Watson
1970 Barbara W. Tuchman	1994 Robert Serber
1971 George E. Reedy	

For my teachers at Columbia College

CONTENTS

Preface

In this book I argue that chemistry is interesting, both to the practitioners of the molecular art, craft, science, and business and to the reflective consumers of its products. The interest derives from an inherent tension. Each fact or process of the science, and the way these are viewed, is in precarious balance between polar extremes. And the polarities of substances and their transformations resonate with forces deep in our psyche.

1.

What do you want when you come to a physician with your aging father, who is weak and feverish? Compassion to be sure, but also a laboratory workup of blood chemistry or a test for the organism possibly causing the suspected pneumonia. And if needed, a drug, an antibiotic tailored to remove that organism from your father's body.

What do I scream about when the town decides to put a huge garbage incinerator, taking in municipal and industrial waste from around the state, next to my home? It's the traffic, the smell, the possible discharges into my well water of certain ions and molecules, and still other pollutants into the air.

The substances you desire from the physician, the substances I worry will turn up in my air and water are chemicals. So are you, and I—chemicals, simple and complex. You certainly want more than a prescription of some chemicals from a doctor—you want care and compassion. And I want more than reassurance and continual monitoring of chemicals emitted from the agency that sites the incinerator—I want fairness, a real consideration of environmental impact, and of alternatives to incineration. But in the material, real world, we—you and I—deal with and react to substances that are chemical.

These chemicals we desire and fear (chemists call them compounds or molecules, once they are reasonably pure) are not the largest (the realm of astronomy), nor the smallest (part of physics). They are squarely, nicely *in the middle,* on our human scale. Which is why we care about them, not as distanced, hypothetical constructs, but in this world. Those molecules, of pharmaceutical or pollutant, are of just the right size to interact, for better or for worse, with the molecules of our bodies.

That a reasonable human being can be ambivalent about chemicals, seeing in them both harm and benefit, is not a sign of irrationality but of humanity. Utility and danger are two poles of a duality. Any fact in our world is evaluated, often subconsciously, by our wonderfully rational and irrational mind, in terms of such polarities. Only if one is dead to experience does one fail to ask the dual question—"Can it help me?" / "Can it hurt me?" Asking that question endows the object of the query, the "it," with a kind of life. It is linked to you. The tension of the object being harmful, or harmless, or maybe both, makes it *interesting.* The etymology of "interest" is from *inter* and *esse,* to be in-between. The tension of asking the question and struggling with the answer links the material and spiritual worlds.

Harm or benefit, harm and benefit, is only one of the polarities that makes chemistry interesting. In this book I will explore others as well. The first will be that of identity. As the title of this book implies, I happen to think this is the most important one. Later on, I will look at dualities such as static/dynamic, creation/discovery, natural/unnatural, and to reveal/to conceal.

A chemical fact—a molecule, a reaction—is poised in some way in the multidimensional real and mental space defined by these dualities. Is it a new molecule or one made before? Is it safe or harmful, and to whom? Is it sitting still, the way it seems to be, or is it really moving at the speed of sound? Is it present in nature or made in the laboratory? Question after question; questions that build tension, especially if the answer is "neither"—or "both." Tension gives life, the potential of change. If there is anything central to chemistry, it is change.

2.

There is a second, connected aim of this book—to tell you what chemists really do. I don't intend to propagandize for chemistry, but to open to you a window into the chemist's world. So that you may see how these dualities, connecting up with psychological forces common to all us, enter the life of the practitioners of the art.

To understand is to give oneself the possibility of not being afraid, perchance be interested. The chemist's world is penetrable. Through case studies I will show you how intellect and tools are marshaled to answer the simple questions anyone would ask: "How do I do it?" "What do I have?" "How did that really happen?" "How shall I tell others, if I am to tell them?" "Is it of value?"

Answering these simple common language questions leads one quite naturally to ponder the dualities underneath. So asking, "What do I have?" becomes "Is this white powder the same or different from a million white powders [yes, there are a million, at least] made previously by others?" I will try to show, by example, how practicing chemists deal with these questions.

3.

Since the theme of polarities I stress bridges both matter and emotion, there is no way to avoid the human person, with his or her immense capacity for curiosity, bold creation, and fear. I will discuss the thalidomide episode, a failure of the system and of individuals. And I will tell of the complicated, creative, and tragic life of a great German chemist, Fritz Haber. I will make a personal statement of what I see as the social responsibility of scientists and an equally personal one on how a chemist might respond to environmental concerns. My aim is for a middle ground, as hard as that may be to find.

4.

Chemists are no more reflective than other people. But the questions they pose, and the craftsmanship with which they answer them, move them to consider polarities, and the associated tensions. Or the dualities press themselves, subconsciously, into the chemist's mind.

The dualities—of molecules and the process of their making—are important, I think, in forming a link between the chemist and the nonchemist. It is possible to answer the question "What do I have?" and to reflect on whether the substance made is the same or not the same as others. But why is that question interesting? Because the question of identity, of *our* identity, shaped in childhood in a complex dance of bonding and separation, matters deeply to us. The processes of nature connect with the interior world of our emotions.

Identity and deception, origins, good and evil, sharing and withholding, resurrection, danger and safety, and overcoming obstacles are some of the psychological constructs or mythical structures with which the world of molecules connects. These emotional focal points shape, consciously and subconsciously, the wonderful, game-playing psychology of the chemist engrossed in molecules. It helps to see this to sense what moves chemists. And I think the material-psychological link, expressed through polarities, allows us to understand why we like and fear chemicals.

THE SAME AND NOT THE SAME

PART ONE

Identity—the Central Problem

1. LIVES OF THE TWINS

Joyce Carol Oates, one of America's most talented and prolific writers, has written several psychological thrillers under a poorly concealed pseudonym, that of Rosamond Smith. In one way or another these novels deal with the complexity, richness, and threat of twinhood, with similarity and difference.

In *Lives of the Twins,* published in 1987, Oates/Smith draws us into the world of a young woman, Molly Marks, who falls in love with her therapist, Jonathan McEwen. It emerges that Jonathan has an identical twin, James, whose existence he has concealed from Molly. Some hidden dark evil has separated the twins. James is also a psychotherapist. Molly, obsessed, seeks out James and begins a complicated relationship with him. Here is Molly's description of the brothers:

> Yes, their hair whorls in opposite directions but it *is* the same hair, precisely—texture, thickness, springiness, degree of silvery-gray streaks and shadings. . . . If their teeth tend to decay on opposite sides of their mouths Molly can't know but, in all, their teeth look very much alike. Each has a slightly jagged left incisor that gives him, to Molly's romantic eye, a rakish razorish air, like Mack the Knife. . . . When Jonathan smokes he holds his cigarette in his right hand, and, exhaling smoke, has a habit of screwing up the right half of his face; James holds his cigarette in his

left hand, and, exhaling a luxurious cloud of smoke, screws up the left side of his face. Jonathan appears to smoke only when he is unhappy while James, who is never, evidently, unhappy, smokes when he pleases. James smokes the brand of cigarettes Jonathan smoked when Molly first knew him; now Jonathan is trying other brands, less potent, and less satisfying, in an effort to stop smoking entirely.

Both brothers use the same brand of razor blades, deodorant, aspirin, toothpaste . . . though James squeezes the toothpaste tube anywhere he wishes while Jonathan squeezes it from the end and neatly rolls it.[1]

What do the sometimes identical, sometimes mirroring habits of a set of fictional twins have to do with chemistry?

Chemistry, the molecular way of knowing the natural and unnatural, is a remarkable science, prodigal in the way it has changed our world. Chemistry touches every aspect of the way that we live—including, for example, James's and Jonathan's use of their razor blades, deodorant, aspirin, and toothpaste. We wear clothes in colors that were once accessible only to potentates; we live when, in an earlier time, we would have died many times over. Illustration 1.1 shows the rate of survival in children afflicted by a variety of solid tumors—plotted as a function of year in this century.[2] Not much happens until chemotherapy is introduced.

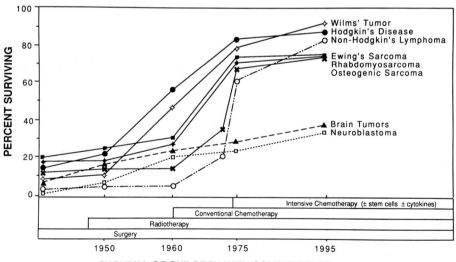

SURVIVAL OF CHILDREN WITH SOLID TUMORS

1.1 Percent survival of children diagnosed with various types of solid tumors over the period 1940–1995. Courtesy of F. Leonard Jones, M.D.

1.2 Waste barrels. (Photo by John Cunningham/Visuals Unlimited)

Through the science of molecules and their transformation we have learned of the invisible inside of matter, the prolific ways in which atoms link up in natural silk and artificial nylon. *And* we wash our apples and other fruits and vegetables not so much because of the dirt on them but because we are afraid of the chemical residues that we ourselves have put on those apples. Illustration 1.2[3] shows a chemical dump site; inefficient industrial production and human failure sometimes combine to pollute our environment.

All this, in the complex beauty of the real world, totally resistant (as human personality is, as art is) to simplistic categorizations of good or evil—all this is chemistry. The Janus image (illustration 1.3[4]) is an apt metaphor for how much of the outside world regards chemistry.

The ambiguity in the way chemistry is perceived is but one, external dichotomy. There is more. Poised centrally between the physical and biological universes, chemistry does not deal with the infinitely small or large, and concerns itself only indirectly with life. So it is sometimes typed as dull in the way things in the middle often are. But there is a dual surprise awaiting the careful observer of the molecular scene; for it is a rich, agitated world down there, both in its innards and in the emotions of the supposedly dispassionate (but actually impassioned)

1.3 Hans Erni, "Janus Image," 1981.

practitioners of molecular arts. In this book the essential tensions of chemistry will be explored; I will seek the polarities that power, rend, and reform the world of molecules.

What *do* twins have to do with it? Everything. The questions implicit in Molly Marks's description of the twins are: "Who are you?"; "Are you different?"; "Are you the same?" The tension for Molly is that of recognition, of identity, of the same and not the same. The same overpowering questions initiate the dialogue of a chemist with recalcitrant matter. He or she also asks, "What are you?"; "Are you different?"; "Are you the same?" The stranger within; the idea of molecular mimicry—these guiding metaphors of immunology and drug design extend the notion of molecular identity. They are strong metaphors, as we will see, because they touch deep concerns of differentiation, of individuation, of the self.

2. WHAT ARE YOU?

The very first question a chemist asks when faced with a sample of anything new under the sun—some dust brought back at fantastic expense from the surface of the moon, an impure narcotic off the street, an elixir extracted from a thousand cockroach glands—is always the same: "What do I have?" This query turns out be more complicated than one might think, for in the real world everything is impure. If you were to look at the purest things in our environment—silicon wafers, table sugar, or some pharmaceuticals—you would find that at the parts-per-million level, you might not *want* to know what is in there!

Everything is in fact quite dirty. Especially natural things, which are much more impure on average than synthetics. Nicely so. Some 900 volatile aroma components have been found in wine;[1] that great German Moselle, the 1976 Bernkasteler Doctor Trockenbeerenauslese, is identified by the expert taster because of the *mix* of ingredients, natural chemicals (what else is there?) which give the wine its taste and smell. Curiously, the taste and smell as a whole, even though the ingredients are chemicals that can be quantified, ultimately elude the chemist's expertise. It takes a wine person, with a discerning palate and a nose, to pick that wine out.

Why are natural things impure? Because living organisms are complex, and they are a product of evolution. You need thousands of chemical reactions, a myriad of chemicals, to "run" a grape or your body. And nature is a tinkerer; the solutions for ensuring survival of a plant or animal are the result of millions of years of random experimentation. The patches on the fabric of life come in a bewildering variety of molecular shapes and colors. Anything that works is co-opted. And banged into shape by all those natural experiments.[2]

So the realistic question becomes not "What is it?" but "How much is there of what?" One must separate a substance into its constituent components. Each component is a *compound,* a persistent grouping of atoms that stick together. That group of atoms is called a *molecule;* a pure compound is a substance consisting of a very large (molecules are tiny) assemblage of identical molecules. Each compound will have quite different properties; sugar and salt may be white crystalline solids soluble in water, but there is no problem in distinguishing them by other physical (and chemical, and biological) attributes.

After separation of a substance into its components, one wants to identify the constituent compounds. To a chemist, *structure* means the identity of the atoms that are in the pure compound, how those atoms are connected to each other, and what their arrangement in space is.

Let us begin with the problem of separation. I happen to be a mineral collector, and illustration 2.1 shows one way nature does it: cubic

2.1 Fluorite on barite. (Photo by Studio Hartmann)

crystals of fluorite, clear to pale lavender, perch on long-bladed crystals of barite in this specimen from the Schwarzwald, Germany's Black Forest region.[3] If you have time (of a geological scale) available to you, then under certain conditions substances might separate from each other naturally, as these did. The method is called fractional crystallization. However, most chemists' patience is not on the order of thousands of years. The five years that a Ph.D. student spends in graduate school is more like it. Human beings want a speedier technique, and so one invents machines to separate things.

Illustration 2.2 is the outcome of such a machine at work. This "gas chromatograph" may cost about $5,000. It separates molecules by a repeated process of adsorbing them on little sandlike grains, then releasing them. In this duality of holding on and letting go, different molecules find a different balance and pass through the machine slower or faster.

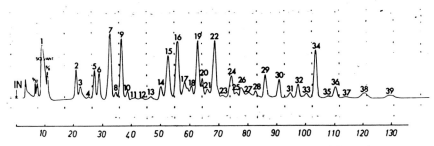

2.2 Thirty-nine peaks, each containing at least one compound from the aroma of cocoa. The horizontal axis is time, in minutes, in which the compound is eluted from a gas chromatograph. The vertical axis is related to the concentration of the components. Reprinted by permission from J. P. Marion et al., *Helvetica Chimica Acta* 50 (1967): 1509–16.

The article from which this illustration is taken describes the work of a group of chemists engaged in analyzing the aroma of fresh cocoa.[4] Why would anyone want to do that? The Nestlé laboratories in Vevey, Switzerland, might well want to do that. Their chemists took a mere two thousand kilograms of Ghanian cocoa, extracted the aroma with steam and dichloromethane. They concentrated the extract to just fifty milliliters. Then they put fractions through the gas chromatograph. In the illustration one can see thirty-nine peaks on some time scale, as they emerge from their ordeal of union and separation in the chromatograph. Every one of those peaks is at least one compound; the Nestlé chemists actually identified fifty-seven different compounds, thirty-five of which had not been previously known to be in cocoa. The complexity of the real world swims out at you. All fifty-seven compounds (each made up of a lot of identical molecules) may not be necessary to give the aroma of cocoa. But it's remarkable how complicated that natural mixture truly is.

The next task is to find out precisely what molecules are in each one of these thirty-nine peaks. In some cases, when the molecules cooperate, if they crystallize neatly, then, with a machine called an X-ray diffractometer (costing about $100,000) and one week's work, it is possible to determine the structure of the molecule.

An example of such a "crystallographically determined" molecular structure is shown in illustration 2.3.[5] This is not a molecule to be found in the aroma of cocoa! It has three rhodium atoms in it—as far as I know no one has found rhodium in cocoa. Not that natural organisms avoid metals; there is a central role for iron, copper, manganese, zinc, magnesium, and even rare molybdenum and selenium in living systems. But rhodium, critical for the operation of your car's catalytic converter (more of this in chapter 34), isn't an essential biological trace element. I show the molecule just to indicate the detail in which one can determine molecular shape. In this "Star Wars" representation you see some numbers; those are distances between the atoms. Even such metrical detail may be gleaned.

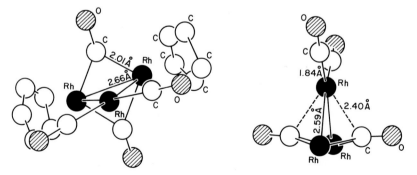

2.3 Two views of the structure (in a crystal) of $Rh_3(C_5H_5)_2(CO)_4^-$.

3. WHIRLIGIGS

Yet often molecules do not cooperate with us by telling their mysteries so directly. They may fail to provide the nice little crystals requisite for the technique (X-ray crystallography) described at the end of the last section. Let me relate a little story of how some chemists determined the structure of a molecule when a direct crystallographic solution was not available. The story is drawn from the work of colleagues of mine—Jerrold Meinwald, an organic chemist, and Thomas Eisner, a neurobiologist, entomologist, and insect physiologist, both at Cornell. They have been working together for the last thirty years on chemical ecology, the defense and communication systems of insects. Insects are the greatest chemists. More than other species, they successfully use simple and complex molecules, singly and in perfumelike blends, to communicate in feeding, defense, reproduction, and behavior.[1]

You can see a typical scene near my hometown of Ithaca in illustration 3.1. It is autumn, our finest season, and maple leaves float on the surface of a pond. On that surface are some little beetles. These interesting organisms, whirligigs, family *Gyrinidae,* live in a unique habitat, on the water's surface. The aspiration of many a fly-fisherman is to

3.1 A pond with whirligigs, in Sapsucker Woods, Ithaca. (Photo by Thomas Eisner, Cornell University)

simulate this scenario. Since whirligigs proliferate, Eisner reasoned that they might have a defense mechanism against predatory fish and amphibians and set out to determine that mechanism.

A gyrinid beetle, *Dineutes hornii,* is shown in detail in illustration 3.2.

3.2 Whirligig, *Dineutes hornii.* (Photo by Thomas Eisner, Cornell University)

From two saclike glands that open near the abdominal tip, the beetles if threatened or mistreated exude a white, milky substance (illustration 3.3). That substance acts as a feeding deterrent to fish and perhaps amphibians.

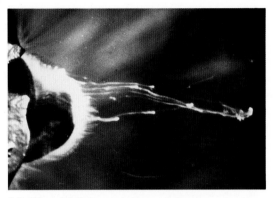

3.3 The whirligig extruding a defensive substance.
(Photo by Thomas Eisner, Cornell University)

From fifty beetles Eisner, Meinwald, and Opheim isolated four milligrams of a yellow oil, named by them "gyrinidal." You can barely see four milligrams; it's a very small drop indeed. But from those four milligrams they were able to determine the structure of the constituent molecules of that vital (to the bug) compound.

How did they do it? The story,[2] very much a detective story, begins with some physical measurements. In illustration 3.4 is the first of several slides Eisner and Meinwald show in describing their chemical work to their professional colleagues:

$$50 \text{ beetles (pygidial glands)} \longrightarrow 4 \text{ mg yellow oil}$$
$$\text{"gyrinidal"}$$

IR: 1680, 1663, 1640, 1618 cm^{-1}

UV: 238 (20,300); 325 (sh.) nm (EtOH)

MS: m/e 234.1254
             234.1256 calcd. for $C_{14}H_{18}O_3$

3.4 Summary of various kinds of spectroscopic observations on gyrinidal, as they might be presented by J. Meinwald in a professional lecture.

I introduce this illustration with some trepidation because it is chock-full of the jargon of the trade. But I can explain roughly what is being measured, why, and what essential information the chemists obtain from these numbers. The details are not unimportant, but the essence of the method is within reach. I think it is worthwhile to risk sampling the reality of modern chemistry through this case study.

As has been noted, molecules are extremely tiny; indeed, if each of the molecules in those four milligrams of gyrinidal oil could be enlarged to just the size of a grain of sand, they would cover Ithaca to a depth of around four hundred meters. No optical microscope can see them. Yet there are ways to probe their structures, called spectroscopies. These poke the molecules with light of one or another color, and the molecular response—with light absorbed or emitted—allows a molecular detective, a chemist, to deduce the structure of the molecule.

Signals from within is what those "spectroscopies" are, and here, phenomenologically, is how it works. You may know that the frequency of a twanged guitar string depends on (a) the length of the vibrating string (that's what frets are for) and (b) the thickness of the string and the material it is made from. The details—exactly what note you get from a brass string one millimeter in diameter—a physicist or engineer can work out. Imagine a strange guitar lying in a room you cannot see into. If you can get someone to pick the guitar, and you know the theory of vibrating strings, from the signals from within that dark room (the guitar playing) you can deduce the length and thickness of the strings.[3]

IR and UV here stand for "infrared" and "ultraviolet" spectroscopies, two machines, two techniques, picking on the light waves and listening for the response. MS is an abbreviation for "mass spectrometry," a third instrument. The cryptic numbers indicate in spectroscopic shorthand the outcome of the measurements, Illustration 3.5 shows one of the machines, a mass spectrometer.

The IR and UV machines cost approximately 5.5 kilodollars apiece while the MS instrument goes for a whopping 220 kilobucks. I stress their costs because someone out there is paying for the games that chemists play. *You* are paying for those games, which bring us fundamental, reliable knowledge of whirligigs and many other aspects of the world. This is research, it is useful, and it is also a "Glass Bead Game."

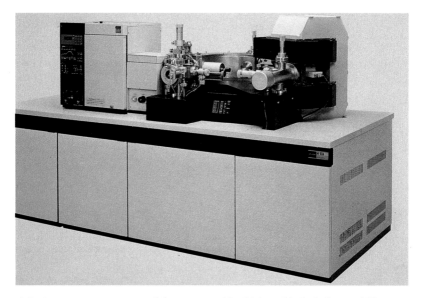

3.5 A mass spectrometer of the type used by Meinwald, Opheim, and Eisner in their study of gyrinidal. The machine portrayed is a Hitachi MS-80A, a contemporary mass spectrometer.

The paymasters should know the sometimes substantial costs of basic research.

The most expensive of these glittering machines, the mass spectrometer, does earn its way. It essentially weighs the molecule and tells us very precisely that in gyrinidal there are fourteen carbons, three oxygens, and precisely eighteen—and not seventeen or nineteen—hydrogens.

But the chemist wants to know more than the $C_{14}H_{18}O_3$ formula. How are those atoms connected to each other, what is the shape of the molecule? The two other spectroscopies used in illustration 3.4 give some clues, but none proves the structure of the molecule. In order to answer that, Meinwald and Opheim next used another machine, costing about two hundred kilobucks, called an NMR spectrometer. What this machine does is to measure the magnetic field at each hydrogen atom that is in this molecule. Every hydrogen atom located in a microscopic environment different from another gives a distinct signal. The little blips you see in illustration 3.6 are once again messages from within, clues to the identity of different hydrogen atoms in gyrinidal. The very same technique is used in MRI, magnetic resonance imaging. And you should most definitely ask how much an MRI installation costs.

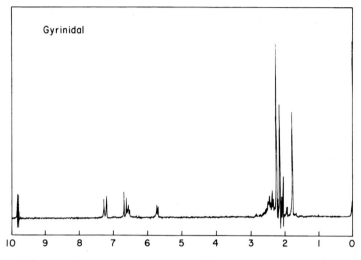

3.6 A nuclear magnetic resonance (NMR) spectrum of gyrinidal.

Here is how the chemists reasoned: In the spectrum (illustration 3.6), there is a peak at about 9.97, another peak at 1.82 and one at 2.27, on some scale. Those peaks, as I said above, are characteristic of hydrogen in different environments. The Cornell crew knew or read that in a thousand other molecules it had been found that whenever there is a peak in this spectrum at about 9.97, that peak is characteristic, a fingerprint of a hydrogen which is bonded to a carbon which in turn has an oxygen connected to it (HCO)—while a peak at 1.82 is associated with a hydrogen that is bound to a carbon that does *not* have an oxygen bonded to it, but instead has two other hydrogens bonded to it (HCH_2). These associations of spectra with structure, and others like them, led the chemists to compose piece by piece the structure of the molecule. In the end, the suggestion (more than that—it is correct!) emerged that gyrinidal is the molecule shown in illustration 3.7.

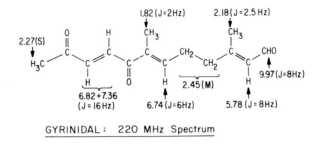

GYRINIDAL : 220 MHz Spectrum

3.7 The deduced structure of gyrinidal. The numbers refer
to the position of the NMR "peaks" shown in illustration 3.6.
This is also a reproduction of a technical slide as used by
J. Meinwald in a lecture on gyrinidal.

This is a structure determination. And a starting point for some reflections. The first thing to be said, one I've hinted at, is that this assign-
ment of atoms and their connections has very much the feel of a detective story. All the precious pieces of evidence, all the blips and peaks
and valleys provided by hundred-thousand-dollar instruments, *none* of
these alone proves anything. They are but clues. In the hands of a
clever person they are assembled and intellectually connected to each
other, like pieces of a puzzle, and very much like a narrative they reveal
to the trained molecular diagnostician the story of a molecule's structure. Most of the time the solution turns out to be right.[4]

4. Fighting Reductionism

The ingenuity that went into assigning a structure to gyrinidal is repeated a thousand times daily by chemists, organic and inorganic. Structure determination uses physical measurements and their interpretation. The chemical practitioner of this art understands roughly the physics behind a given spectroscopy, but then uses that physics often by analogy, observing that a thousand other compounds have such and such a peak at a certain place in the spectrum. To some people that does not constitute sufficient understanding. They would say that you need to delve deeper into the physics, identify the various mechanisms or causes which are behind that signal from within, and actually compute their outcome. You should not claim to *understand* the technique until you really know that a peak should come at 9.97 and not, for example, 9.87 or 10.07.

What can one say about a person who seeks that kind of understanding? We cannot deny that it is a good thing. That searcher for understanding will go back deeper and deeper, entering a reductionist mode. He or she will become engrossed in the sources of the physical phenomenon, likely do good science. But I hazard the guess that he or she will not solve many structures. The psychology of finding solutions involves a certain mental "drawing of a line," a self-imposed limit on how deep you need to go in. The people who go deeper and deeper are seeking another kind of knowledge than those who want to solve the problem.

This brings us to reductionism and ways of understanding. By reductionism I mean the idea that there is a hierarchy of sciences, with an associated definition of understanding and an implied value judgment about the quality of that understanding. That hierarchy goes from the humanities, through the social sciences to biology, to chemistry, physics, and mathematics. In a caricature of reductionism one aspires to the day when literature and the social sciences will be explained by biological functions, biological ones by chemical ones, and so on. We probably owe the beginnings of this philosophy to Descartes, and its more explicit statement to Auguste Comte and the French rationalist tradition.[1]

Scientists have bought the reductionist mode of thinking as their guiding ideology. Yet this philosophy bears so little relationship to the reality within which scientists themselves operate. And it carries potential danger to the discourse of scientists with the rest of society.

I think the reality of understanding is the following: Every field of human knowledge or art develops its own complexity of questions. The problems facing chemistry are in some ways more complex than the ones in physics. Much of what people call understanding is a discussion of questions in the context of the complexity or hierarchy of concepts which are developed within that field. If you wanted to deprecate this way of thinking, you would call it quasi-circular. I wouldn't deprecate it; I think this kind of understanding is quintessentially human and has led to great art and science.[2]

There are vertical and horizontal ways of understanding. The vertical way is by reducing a phenomenon to something deeper—classical reductionism. The horizontal way is by analyzing the phenomenon within its own discipline and seeing its relationships to other concepts of equal complexity.

Let me illustrate the futility of reductionism with a reductio ad absurdum. Suppose you receive an anonymous letter. In that letter is a sheet of paper with a four-line poem, "Eternity," by William Blake:

> He who binds to himself a joy
> Does the wingéd life destroy
> But he who kisses the joy as it flies
> Lives in eternity's sun rise.

Knowing the sequence of firing of neurons when the poet wrote that line, or in your mind when you read it, or in the mind of the person who sent the letter, knowing the fantastic, beautiful complexity of biochemical actions behind the firing of neurons and the physics and chemistry behind that, that knowledge *is* incredible and desirable, that knowledge is going to get you a lot of Nobel Prizes, I want that knowledge, *but* . . . it has nothing to do with understanding the poem, in the sense that you and I understand a poem, or drive a car, or otherwise live in this terrible and wonderful world. The "understanding" of Blake's poem is to be sought at the level of the language in which it is written, and the psychology involved in the writing and reading of it. Not in the firing of neurons.

If you are willing to accept a leap between the humanities and science, I tell you that even in two "hard natural science" fields as close to each other as chemistry and physics, even *there* there are concepts in chemistry which are not reducible to physics. Or if they are so reduced, they lose much that is interesting about them. I would ask the reader who is a chemist to think of ideas such as aromaticity, acidity and basicity, the concept of a functional group, or a substituent effect. Those constructs have a tendency to wilt at the edges as one tries to define them too closely. They cannot be mathematicized, they cannot be defined unambiguously, but they are of fantastic utility to our science.[3]

Reductionism is often used as a psychological crutch rather than a realistic description of how understanding functions. You might think, for instance, that physicists would be happy with a reductionist philosophy, because they are near the base. Deeper still, perhaps, are mathematicians. One might expect therefore that physicists should have a positive attitude toward mathematicians. But just ask your local physicist what his or her feeling is about mathematicians. What you usually get is a host of negative responses such as "mathematicians are impractical," "they do not take their inspiration from us," "they do not deal with reality." It is obvious that for physicists the reductionist chain stops at physics. And for a chemist, talking to an economist or biologist, it too often stops at chemistry.

Moreover, adherence to reductionist philosophy is potentially dangerous. A vertical mode of understanding, if championed as the *only* mode of understanding, creates a gap between us and our friends in the arts and humanities. They know very well that there isn't just one way of "understanding" or dealing with the death of a parent, or our country's drug problem, or a woodcut by Ernst Ludwig Kirchner. The world out there is refractory to reduction, and if we insist that it must be reducible, all that we do is to put ourselves into a box. The box is the limited class of problems that are susceptible to a reductionist understanding. It's a small box.[4]

5. The Fish, the Worm, and the Molecule

After that minor tirade against reductionism, let me return to what Meinwald, Opheim, and Eisner did. They assigned the structure of gyrinidal, as we have seen. Then they synthesized it, in a laboratory. They actually *made* gyrinidal. I will return to synthesis, the contender with analysis for the heart of chemistry. Here let me show the Cornell researchers' bioassay, a check whether the synthetic material be active or not.

In illustration 5.1, the picture at upper left shows a hungry largemouth bass that has not eaten for several days. In the same picture is a worm, painted with four-tenths of a microgram of the synthetic material (that's precious little; without a microscope you could not actually see four-tenths of a microgram). The fish does what it has been programmed to do (illustration 5.1, upper right). You can see the worm in its mouth. The bass then goes through a form of instinctive behavior by rinsing the unpleasant tasting object (illustration 5.1, lower left), eventually deciding (illustration 5.1, lower right) that the worm was not for it.

Does that prove anything? No, it does not. It does not prove that the synthetic material is identical with the natural one. Other tests accom-

5.1 The bioassay of synthetic gyrinidal. (Photos by Thomas Eisner, Cornell University)

plish that. But the bioassay does prove that the synthetic material is a good feeding deterrent, an interesting and potentially useful piece of information in the never-ending quest for reliable knowledge. And while not proving the identity of the natural and synthetic material, it does bolster the confidence of the investigators that they are identical. That is not an unimportant psychological role for circumstantial evidence; the task of searching for reliable knowledge is difficult, even wearisome. One needs all the support one can get. And not just a research grant.

Even though the primary chemical activity under discussion here, synthesis, will be taken up in more detail later on, I cannot forgo quoting here a story of another bioassay, as told by John Cornforth, a great organic chemist:

> A team in the United States was studying the sex attractant of the female American cockroach. Warm air was passed over a very large number of

these animals and then through a cold trap. By further refining a minute amount of active material was obtained and a structure was proposed on the basis of physical measurements (illustration 5.2).

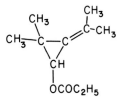

5.2 The proposed (but incorrect) structure of the American cockroach sex attractant.

There were then, as now, a considerable number of chemists looking hungrily for an excuse to synthesize something, and the effect of this structure was rather like that of a dead horse dropped into a lake of piranha fish. Here was a small molecule asking for the application of up-to-date reactions, and the excuse for synthesizing it was most grant-worthy: an adequate supply of the stuff, obviously not available from nat-ural sources, might plausibly play its part in controlling a noxious pest. Within 3 years, six approaches were reported, all most ingenious. Two of them were successful, the others were honourable near misses. So the molecule was well and truly synthesized and the compound became readily available. There was only one snag—the proposed structure was wrong and the synthetic material inactive. A lady I know remarked at the time that, although this molecule wasn't very good at attracting male cockroaches, it certainly attracted a lot of organic chemists . . . but per-haps it would be kinder to say that synthesis here was the final proof of non-structure.[1]

6. Telling Them Apart

Back to Rosamond Smith's *Lives of the Twins*. The souls of James and Jonathan are as different as can be. Or are they? Molly is trapped in her love for both of them. Late in the book, in the critical moment when for the first time the twins and she meet, Smith writes:

> One of the men is running toward Molly; the other begins to run too, and overtakes him. Tall heavy-headed men, broad-shouldered, with dark hair: identical-appearing, or—are they in fact identical? Molly has never seen the men together before and has never before experienced this visceral shock, a dread powerful as a kick in the belly. She recalls the fear of twins, in primitive cultures; the superstitions about twins—that one, or both should be killed at birth. How can such creatures be told apart? How, without their cooperation and consent?[1]

How indeed can creatures be told apart, without their cooperation and consent? The basic tension in chemistry, one there from the beginning, is that of the twins, of the same and not the same, of identity, of the self and not-self. Along with other dualities, which we will explore, this one propels the science. Could it be that it does so because it touches upon something deep in our psyche?

7. Isomerism

Let me be more specific about the problem of identity in chemistry. We have learned, with much ingenuity and effort, that all matter is made of molecules, which in turn are composed of atoms. Some matter is atomic in its composition (helium or argon gas), some made all of one element but the atoms are linked up in some simple or complex way (the iron atoms in iron metal, carbon in graphite or diamond). But most things are molecular, made up of persistent groupings of bonded atoms.[1]

That icon of chemistry, Mendeleyev's Periodic Table of the Elements, is shown in illustration 7.1. There are about ninety natural elements, fifteen or so radioactive, man- or woman-made ones. But what a dull world it would be if there were only 105 things in it! Any square foot of this beautiful world shows a far greater richness. Everything in the world—whether it be sugar, aspirin, DNA, bronze, hemoglobin— is made of molecules, molecules with reproducible colors, chemical

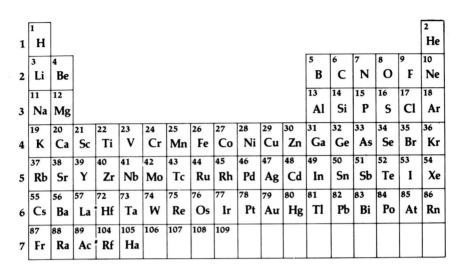

7.1 The Periodic Table. Reproduced by permission from A. J. Harrison and E. S. Weaver, *Chemistry* (New York: Harcourt Brace Jovanovich, 1991), p. 110.

properties, toxicity, all of which are a consequence not only of the identity of their atomic components but also of the way those atoms are connected to each other.

That connection between atoms is called a bond. And boy, do they bind! It's not, however, a random coupling; there be rules to this cross between a donnybrook and a love affair. So carbon typically binds to four others, and hydrogen forms a liaison (indeed that's the French word for bond) with one other atom. And then the game is on between the two, for one does not have CH—at least not much of it (that wouldn't satisfy carbon's constrained lust for bonding, and when CH is found it is a most reactive unstable species)—but CH_4, methane. One can also form carbon-carbon bonds, and the constructive game begins in earnest, with the hydrocarbon series: methane, ethane, propane, and so on (illustration 7.2). The chain builds; its approach to infinity is that ubiquitous polymer, *the* most important plastic of our times, polyethylene (illustration 7.3).

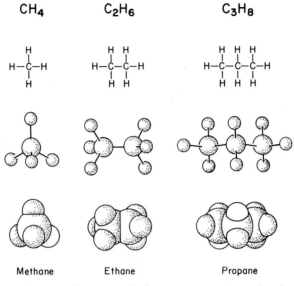

7.2 Methane, ethane, and propane in three representations: the chemical structure, a ball-and-stick model, and a space-filling model.

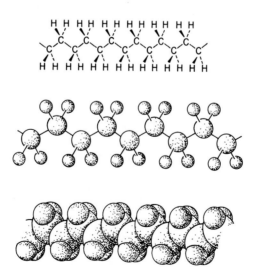

7.3 Polyethylene $(CH_2)_n$, in three representations.

Very soon one finds that the rules of the game (very simple—each carbon can form four bonds, each hydrogen one) allow two or more molecules made up of the same atoms, containing the same number and type of bonds, to exist. Thus for C_4H_{10} we have *n*-butane and *iso*-butane (illustration 7.4):

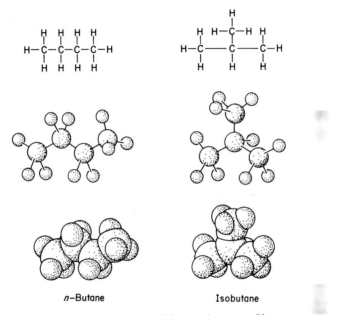

n-Butane   Isobutane

7.4 Several representations of the two isomers of butane.

Each has three C-C bonds and ten C-H bonds. Yet they are different; not very different, mind you, but different enough—in their volatility, in the heat generated when these petroleum constituents burn—to *matter.*

The phenomenon is called *isomerism,* its elucidation a triumph of nineteenth-century chemistry. As the number of atoms grows, the possibilities for isomerism increase. There are two butanes, but three *structural isomers* of pentane, the five-carbon hydrocarbon (illustration 7.5):

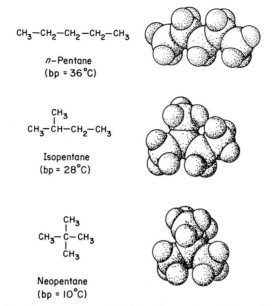

$$CH_3-CH_2-CH_2-CH_2-CH_3$$

n-Pentane
(bp = 36°C)

$$CH_3-\overset{\overset{\displaystyle CH_3}{|}}{CH}-CH_2-CH_3$$

Isopentane
(bp = 28°C)

$$CH_3-\overset{\overset{\displaystyle CH_3}{|}}{\underset{\underset{\displaystyle CH_3}{|}}{C}}-CH_3$$

Neopentane
(bp = 10°C)

7.5 The three pentanes. Their boiling points differ, as indicated.

The number of isomers grows with a vengeance, as this table for the alkanes (a class of compounds of carbon and hydrogen) reveals.[2] And the molecules of life are hardly as small as these. The chemical formula of hemoglobin (a molecule to which we will return several times) is $C_{2954}H_{4516}N_{780}O_{806}S_{12}Fe_4$. Think about the number of isomers of this natural molecule!

Table 1
Number of Structural Isomers of the Hydrocarbons C_nH_{2n+2}

Formula	Number of Isomers
CH_4	1
C_2H_6	1
C_3H_8	1
C_4H_{10}	2
C_5H_{12}	3
C_6H_{14}	5
C_7H_{16}	9
C_8H_{18}	18
C_9H_{20}	35
$C_{10}H_{22}$	75
$C_{15}H_{32}$	4,347
$C_{20}H_{42}$	366,319
$C_{30}H_{62}$	4,111,846,763
$C_{40}H_{82}$	62,491,178,805,831

Before we curse the perverse complexity of nature, we need to relax, to stop and realize (as I said above) that the multiplicity of function of something as elaborate as a human body *requires* such complexity. And more. Diversification provides richness; simplicity might be comfortable for our weak minds but not for the living text of this world.

Structural isomerism is not the only type of isomerism one has. There is also geometrical isomerism, exemplified by the two ethylenes substituted by two bromines, as shown in illustration 7.6. Note that in both $C_2H_2Br_2$ isomers the atoms are connected up in the same way, but that there is a difference in geometry—in one case the two bromines are next to each other, in the other case they are opposite.

cis-1,2-dibromoethylene   *trans*-1,2-dibromoethylene

7.6 *Cis* and *trans*-1,2-dibromoethylenes, geometrical isomers.

In the visual system, in the cones and rods of our retina, the energy of the light causes a transformation of one geometrical isomer of a molecule called retinal into another. The change is not that different in its essence from that illustrated for dibromoethylene. A nerve impulse is triggered, and eventually the molecule returns to the original geometry, ready for the next photon.

Still another place where *cis* and *trans* matters is for our infatuation (no pun intended) with fats. These are long chain molecules which look like a piece of polyethylene with one or more double bonds in the middle of the chain, and a COOH (acid) group at the end. *Cis* and *trans* arrangements at the double bonds are possible; the *cis* isomers are kinked in appearance, the *trans* ones more linear. These geometrical differences translate into biological reactivity. The *trans* fatty acids increase the undesired low-density lipoprotein cholesterol (LDL) in blood and decrease the "good" cholesterol (HDL). This is why "*trans*-unsaturated" (as well as saturated) fats are not particularly good for us.[3]

Little geometrical details matter—the body cares.

8. Are There Two Identical Molecules?

Here are these tiny entities, which we've learned to know and (some of us) to love. In a slurp of water there is this incredibly large number (about 10^{24}) of water molecules. And they're all identical, aren't they?

Not quite. We have to worry about *isotopes*. These are modifications of the atom of an element, in which the nucleus is different (a different number of neutrons in there, along with the same number of protons) but the number of electrons is the same. So for hydrogen there are three isotopes: normal hydrogen (one electron moving around a nucleus made up of a single proton); "heavy" hydrogen or deuterium (one electron around a nucleus containing one proton and one neutron), and tritium (one electron still, but now a nucleus with one proton, two neutrons). In the official nomenclature of isotopes, the total number of protons and neutrons together is given as a superscript preceding the symbol for the element.

$^1H = H$	$^2H = D$	$^3H = T$
hydrogen	deuterium	tritium

Since the mass of the atom resides predominantly in the protons and neutrons, there is a difference in the weight of these isotopes—an atom of deuterium weighs roughly twice as much as one of normal hydrogen (ergo, "heavy" hydrogen) and tritium three times as much. Tritium is radioactive to boot; its nucleus falls apart spontaneously.

The mass of isotopes makes them different. But are they different chemically? The question is not silly—there is difference and difference. Clint Eastwood and Woody Allen are certainly different as personalities. But to a surgeon operating on them and looking for where their aorta might be relative to their hearts, they might not appear to be that different.

Chemistry is *not* determined (in a first approximation) by the nature of atomic nuclei. The energies in chemical reactions are, thank God, nowhere near the energies necessary to initiate nuclear reactions. Chemistry is controlled by the electrons of atoms—the daily miracles of hemoglobin binding oxygen or of your gas stove lighting come from the outer regions of atoms, where the electrons roam, "feeling" each other. What makes hydrogen act "chemically" as hydrogen is the number of electrons around the nucleus, which matches the number of protons but doesn't depend on the neutron count.

This is why the molecules made of elements that exist in a mixture of isotopes are a wonderful example of the same and not the same. The isotopic modifications of a molecule are different enough so that we can tell they are there (with an instrument costing a few kilodollars, a cheap version of that mass spectrometer mentioned earlier, we can weigh them). But not different enough to matter, meaning their chemistry is nearly the same.[1]

Let me be specific about water. On earth, the natural abundances of H, D, T, and the three naturally occurring oxygen isotopes ^{16}O, ^{17}O, ^{18}O are given below:[2]

H	99.985%	^{16}O	99.759%
D	0.015%	^{17}O	0.037%
T	10^{-20}%	^{18}O	0.204%

Now where did these come from? The isotope ratios were set by the nuclear burning processes in the first minutes of the formation of the universe, and by the specific history of the formation of our solar system and planet. They would be a little different on a planet of a sun in a distant galaxy. Isotope abundances are a terrestrial given, though there is a small geographic variation in them. The lifetime of radioactive tritium is so short (the half-life of tritium is twelve years) that none of it is around from the beginning; it is all created, quite naturally, by the cosmic rays impacting on the earth.

So there isn't one naturally occurring water molecule, but a denumerable number of them—to be precise, eighteen kinds. Six are drawn below, there also six containing ^{17}O, and another six with ^{18}O (illustration 8.1).

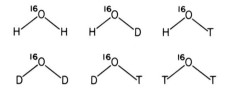

8.1   The six isotopomers of H_2O with the oxygen isotope of atomic mass 16.

It is an easy matter to calculate the relative abundances of these *isotopomers*, isomers only as a consequence of the isotopic difference of their component atoms. $H_2^{16}O$ is the most common, 99.8 times as likely to be found in that slurp as $H_2^{18}O$. And $T_2^{17}O$ is the least common; on the average, not a single molecule in the slurp, or even on earth, will be of this kind.

All of these are natural, all are water. It would be very unhealthy to drink pure T_2O, not because of its chemistry but because of its radioactivity. The little tritiated water in normal water is something we have evolved with over millions of years. It may even be that the chance variation provided by mutations induced in part by this radioactivity was necessary to get us to the present stage of human creative complexity.

Are there two identical water molecules? Sure. In that slurp of water, 99.8 percent of the 10^{24} water molecules are identical. That's a lot of sameness.

But water is simple. Let's get into a living organism and consider a protein, hemoglobin. It contains lots of atoms: to be precise, 2,954 carbons, 4,516 hydrogens, 780 nitrogens, 806 oxygens, 12 sulfurs, and 4 irons. Carbon exists naturally in three isotopes: ^{12}C (most abundant), ^{13}C, ^{14}C. So do hydrogen and oxygen, as we have seen. Nitrogen comes in two naturally occurring isotopic forms, sulfur in four, iron in four as well. (There are four iron atoms per hemoglobin; they're essential in the activity of the molecule.) The number of different isotopomers of hemoglobin is astronomical (whoops, why not just call it chemical!). Working out the combinatorics, one comes to the conclusion that for such a large molecule, even with an immense number of molecules (about 10^{17} molecules of hemoglobin in a drop of blood), that the chances of two tiny hemoglobin molecules fished out of that drop being exactly the same, in every isotopic detail, are very, very small! Henning Hopf, who suggested that I discuss this subject, speaks of this as the "individualization of compounds."[3]

So the answer to the question heading this chapter is "No, for a really large molecule, probably there are no two identical molecules in that Burmese cat." But does it matter chemically, or biologically? No, the chemistry (and the benefit or toxicity to humans) of all those tiny entities differing only in isotopic composition is nearly identical. They are different, but not different enough to matter—like the maple leaves off the trees in my yard, when all I want to do is rake them up.

9. Handshakes in the Dark

In the realm of differences, still more subtle is chirality, or handedness (from *cheiros,* the Greek word for "hand"). Some molecules exist in distinct mirror-image forms, related to each other as a left hand is to a right. Many, but not all of the macroscopic properties of the compounds made up of such mirror-image molecules are the same—they have identical melting points, colors, and so on. But some properties differ, often critically so. This is, for instance, true of their interaction with other handed molecules, such as the ones we have in our bodies. So the *enantiomers* (for that is the name for the distinct handed forms of a chiral molecule) may have drastically different biological properties. One may taste sweet while its mirror form is tasteless. And the mirror-image form of morphine is a much less potent pain reliever.

Our knowledge of chirality begins in 1850 with a 26-year-old Louis Pasteur, before he studied microorganisms or invented pasteurization or developed a vaccine for rabies. He became interested in optical rotation and linked it to a curious problem of the nonidentity of two compounds that should have been identical.[1]

It is in the nature of nature that its fuzzy details and seeming obscurities are clues to the world within. Optical rotation appears a curiosity even today, dealing as it does with the ability of some substances to

This essay is adapted from "A Hands-on Approach" in Roald Hoffmann and Vivian Torrence, *Chemistry Imagined* (Washington, D.C.: Smithsonian Institution Press, 1993), pp. 95–99.

rotate the plane of polarized light—a discovery made in France in the early nineteenth century. Light is a wave. With that wave there move along electric and magnetic fields, oscillating in space and time. In normal light the wavelike oscillation takes place in every plane. But it is possible to filter out "plane-polarized" light, which is still light, still possessing color and intensity, but different in the following way—in plane-polarized light the electric and magnetic fields that make up the light are restricted to oscillating in only one plane. The filters that create such special light from ordinary light are called polarizers. You've seen them in sunglasses and some airplane windows, and the Polaroid Corporation has made a lot of money on them.

Chemical compounds that rotate the plane of polarized light were discovered. You put in light polarized in one plane, and after passing through the crystal it is polarized in another, different plane. And French scientists observed that solid quartz crystals that were mirror images in their external appearance rotated the plane of polarized light in opposite directions.

Meanwhile, a puzzle arose in the chemistry of another important part of French culture—winemaking. You've probably seen fine, colorless crystals, perhaps growing on the cork, in some white wines. These are a product of winemaking (and they are much more abundant on the inside of wine casks and fermentation vessels!), a salt of tartaric acid. That naturally occurring material is in fact optically active, as are most biological molecules. In another stage of the fermentation process, a substance, racemic acid, was isolated. It had precisely the same atomic composition as tartaric acid. But racemic acid did not rotate the plane of polarized light; it was optically inactive. The same—but not the same.

Pasteur recrystallized a salt of racemic acid. Looking at the crystals under the microscope, he noted that they came in two very similar but nonsuperimposable varieties. Painstakingly, with tweezers, he separated the mirror-image crystal shapes, left-handed to one side, right-handed to the other. When dissolved, the solution of the two crystal forms rotated the plane of polarized light in opposite directions—one clockwise, the other counterclockwise. And one was identical to the naturally occurring tartaric acid.

Racemic acid is a 1:1 mixture of optically active tartaric acid and its mirror-image enantiomer. These substances were not only differentiated in their crystal form (which we now know was a lucky accident; most often mixtures of left- and right-handed forms of a molecule co-crystallize in a form that is not handed). They are also optically active in solution; this must mean that the handedness is there not only in the big crystals but resides deeper, in the tiny molecules there in solution.

One of the dramatic moments in the history of chemistry occurred when the dean of French optical rotation studies, Jean Baptiste Biot, skeptical of Pasteur's report, summoned Pasteur to repeat his experiment in his laboratory. Biot prepared the salt of racemic acid according to Pasteur's prescription, Pasteur separated the crystals under the microscope, right there, under Biot's eyes. Biot dissolved the small samples of segregated crystals and measured their optical rotation himself. Such is the essence of reliable knowledge—a reproducible experiment!

It took a quarter of a century, and the work of two other young chemists in their twenties, J. H. van't Hoff in Leyden and J. A. Le Bel in Strasbourg, to explain, in molecular detail, what is behind optical activity. They proposed that carbon atoms are "tetrahedral," meaning that the four bonds carbon forms are along the directions of a regular tetrahedron (illustration 9.1):

9.1 A tetrahedral carbon atom.

Our notation here is that standard (primitive) visual code chemists use for describing three-dimensional structures: a solid line is in the plane of the paper, a dashed line points "in back" of that plane, a wedge to the front.

Now consider the possible existence and identity of mirror-image forms, given the tetrahedral geometry of carbon. If you have one, or two, or three different substituents around a carbon atom (and that's what synthetic chemistry is about, changing one piece of a molecule for another), then the mirror image is identical to the molecule mirrored. Not so for *four* different groups on carbon, as shown in illustration 9.2:

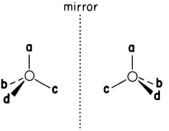

9.2 Nonsuperimposable mirror images.

The molecule at left is *not* identical to the one at right. The only way to convince yourself of that fact is to try to superimpose the mirror images. If you put **a** and **b** on top of each other, **c** and **d** will be out of place. If you superimpose **a** and **d, b** and **c** won't fit. Illustration 9.3 shows the unavailing attempt to superimpose such nonsuperimposable mirror-image molecules (called enantiomers, as has been noted). Molecules that have the potential of existing as left- and right-handed forms are called *chiral.*

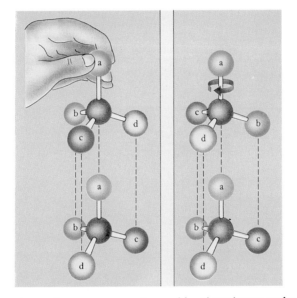

9.3 Trying to superimpose nonsuperimposable mirror-image molecules. Reproduced with permission from J. D. Joesten, D. O. Johnston, J. T. Netterville, and J. L. Wood, *World of Chemistry* (Philadelphia: Saunders, 1991), p. 363.

What does all this have to do with hands? This may not be at first obvious, but the essential descriptors of a hand are thumb, pinkie, palm, and back. They play exactly the same role as **a, b, c,** and **d,** the chemical groups that differentiate enantiomers. There is much more detail to a hand (fingerprints, your lifeline), and so there is to a molecule. But the topological essence of a hand or a molecule with one carbon is described by four markers.

How does one separate mirror-image molecules from each other? The sorting of crystals, which is what Pasteur did, doesn't work very often. Another method, also devised by Pasteur, is to feed a living organism the mixture of enantiomers. The bacteria typically metabolize a molecule of one handedness, excrete the other. But there are many molecules even a bacterium won't eat. Here is scenario from an unmade Antonioni film illustrating the most common method of what is called "optical resolution":

> You (a right-hander) are about to enter a pitch-black room filled with mannequin parts, left and right hands. If you are not able to separate them, something terrible will happen to you. No problem. You begin to shake hands with the myriad paste mannequin hands. You put to one side the ones you can shake hands with comfortably, to the other side the misfit left hands.

In resolution, a handed reagent is added to the mixture of left- and right-handed molecules. It forms two physically distinct compounds—

9.4 Raphael's
Sistine Madonna.
Which is the
mirror image?

the composite of a right hand shaking a left mannequin hand is different in shape, and not the mirror image of, a right hand shaking a right hand. These are separable, they have different properties. They are separated, and then, in a subsequent reaction, knocked apart to yield the components.

Telling left from right is no trivial matter. Heinrich Wölfflin, the great art historian, recounts how every lecturer in art has suffered from the incorrectly inserted slide. The mirror image comes on the screen and one instinctively says, "That is not right!" Wölfflin goes on to say that we should really stop and ask why we think an image that is devoid of obvious gaffes, such as inverted writing, or an army of left-handed swordbearers, why *seemingly* left-right neutral images should be perceived as correct or incorrect.[2]

From a detailed discussion of mainly European examples (see, e.g., illustration 9.4) Wölfflin argues cogently for a shared code between

artist and viewer, a psychologically deep and culturally conditioned way to read paintings.

Left and right *do* matter in biology because our bodies are molecular and handed. Our proteins are like the hands in the dark room scenario; usually they respond differently to handed molecules. Illustration 9.5 shows d- and l-carvone, two enantiomers, in two- and three-dimensional representations (d-carvone can be isolated from caraway and dill seed, l-carvone from spearmint). And they are responsible for a good part of the taste and odor of these plants—they smell like caraway or spearmint—whether they are the natural extracts or are made in the laboratory.

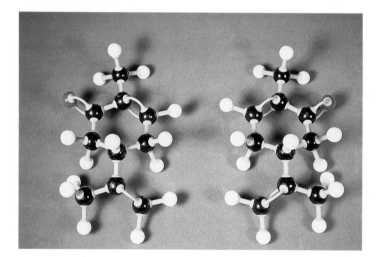

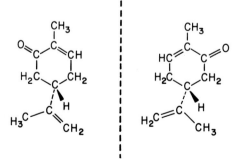

9.5 d- and l-carvone represented by ball-and-stick models (*top*) and structural formulas (*bottom*). (Photo of model courtesy of David N. Harpp, McGill University)

That enantiomers may smell differently tells us that the human smell receptors are chiral molecules themselves, like a left- or right-handed glove. Our receptors can discriminate between left and right hands.

Samples of the enantiomers (illustration 9.6, left) don't *look* very different. But to our nose, or to polarized light, they *are* different. And we use them—in spices or in toothpaste, or to flavor chewing gum (illustration 9.6, right).[3]

9.6 d- and l-carvone as "chemicals" and in natural and unnatural products. (Photos courtesy of David N. Harpp, McGill University)

There are chiral pharmaceuticals whose curative properties are due to just one of the two handed forms and others where both enantiomers are effective but one is toxic or harmful in some way. One such drug is D-penicillamine, widely used in rheumatic arthritis.[4] I'll tell the sad story of another handed drug, thalidomide, in chapter 27.[5]

10. Molecular Mimicry

So there is complexity aplenty in the molecular world—and with that delightful richness the problem of telling molecules apart from each other. How do we distinguish A from B, the left hand from the right? Or friend from foe, self from nonself? Recognition, and its intended subversion, deception, are critical; this is the molecular modus operandi of many poisons in our bodies, and it is the way by which many pharmaceuticals aid us.

Well, why not? Mimicry worked for Jacob, who with his mother's connivance used aroma and touch to fool Isaac and cheat Esau of his paternal blessing; and another deception led to the fall of Troy (both actions being of positive or negative value, depending on your perspective—and critical for history, independent of that perspective). Here are four little tales of chemical deception, natural and unnatural.

1.

Hemoglobin, a marvelous protein, carries oxygen from our lungs to our cells. Illustration 10.1 is a schematic picture of this quite incredible molecule, which, incidentally, we probably know in greater detail than any other biological molecule of like complexity.[1] Hemoglobin is assembled from four tightly fitting chunks, or "subunits." Remarkably, these actually change twice in the course of fetal development, so as to optimize oxygen uptake.

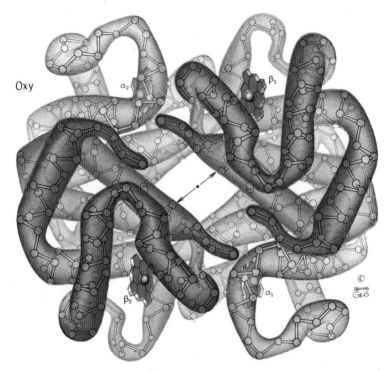

10.1   Hemoglobin in its oxygenated form. (Drawing © by Irving Geis)

Each of the oxygen carrier's subunits is a curled-up chain of atoms (a "polypeptide" built of amino acids) about 440 atoms long. Nestled in the arms of this involuted molecular masterpiece of functional complexity is a platelet-shaped molecular unit, a *heme*. At its center is an iron atom, and it is to this iron that O_2, the oxygen, binds. In each subunit the folds of the protein around it shape a pocket (even equipped with a molecular gate), which guides the oxygen in and out and helps bind it to the iron atom.

The evolutionary tinkering that led to hemoglobin apparently took place in the absence of much carbon monoxide. Then we, humankind, came along, and incomplete combustions occasioned by us and our tools (notably, the automobile engine) now may generate locally high concentrations of CO. The carbon monoxide enters hemoglobin's pocket—so exquisitely crafted by the workings of chance for O_2—and binds to the iron of heme a couple of hundred times better than the intended O_2. The hemoglobin molecules are used up by the CO and are incapacitated for their function. The cells begin to starve for oxygen.

In this instance, binding something else better than what was "intended" can kill, at least potentially. To call this "deception" is a bit of anthropomorphic excess, for the carbon monoxide has no will. The molecule does what it does—the chemistry just works better for CO. Nevertheless, this is a story of similarities and differences—the small size of CO and its likeness to O_2 allows it to enter that protein pocket. In chapter 35 we shall see how one may devise a catalyst for reducing the CO emitted from the internal combustion engine's exhaust.

2.

Competitive binding may be used to save as well as kill. Ethylene glycol is an effective antifreeze that is occasionally ingested, by accident or design. The ethylene glycol is not of itself toxic, but it is converted by a series of enzymes in our body to oxalic acid (present also in raw rhubarb leaves), which damages the kidneys.

In the body of someone who accidentally drinks antifreeze, the first enzyme that gets its molecular hands on ethylene glycol is alcohol dehydrogenase. An enzyme is a small chemical factory, a protein that efficiently catalyzes some chemical reaction. Enzymes are named prosaically after the task that they accomplish, and alcohol dehydrogenase is used by the body to take off, as its name implies, some hydrogen from a class of molecules called alcohols. Ethylene glycol is one of these. Common alcohol, ethanol, is another.

A therapy for ethylene glycol poisoning consists of administering nearly intoxicating doses of ethanol. The ethanol competes effectively for alcohol dehydrogenase's loving arms and so blocks the transformation of ethylene glycol, which is then excreted untransformed.[2] The two molecules, ethylene glycol and ethanol, are shown in illustration 10.2.

10.2 Ethylene glycol (*left*) and ethanol (*right*).

They are indeed similar, but the similarity is actually more of chemical function than of shape. Both are alcohols, which chemically means that they contain an OH group. Such sets of atoms carry with them a characteristic set of properties, such as color and reactivity. Other "functional groups" you may have heard of are COOH (organic acid), HCO (aldehyde), CN (cyanide), ROR (ether, R is any organic group). Functional groups on a molecule are like differently shaped handles on a pot, allowing ways of recognition that are needed by molecules colliding madly (figuratively if not literally in the dark) in a test tube.

3.

Chemotherapy, the use of synthetic chemicals in the treatment of disease, began in 1909 with Paul Ehrlich's discovery of Salvarsan (arsphenamine) for the treatment of syphilis. The great development of the German dye industry—with its structured collaboration of proficient chemists (making hundreds of new compounds) with biologists, pharmacists, and physicians (who then systematically tested these potential drugs in animals and humans)—led to a series of early successes for protozoan and tropical diseases. (Only a half century later, the signal failure to perform such tests adequately caused the thalidomide disaster, a terrible story I will tell later in this book.)

Despite these early successes, however, there was no effective antibacterial agent until the mid-1930s and the appearance of the sulfa drugs. The molecules that eventually became the sulfa drugs were synthesized by the German chemical conglomerate, IG Farbenindustrie, initially because of their potential use as dyes. The first, p-aminobenzenesulfonamide or sulfanilamide, was made in 1908. There were hints that it and related dyes were bactericidal, but these were not pursued systematically until 1932–35, when Gerhard Domagk, a director of research for experimental pathology and bacteriology at IG Farben, undertook a careful study of the biological activity of this molecule. He was aided by the capabilities of the chemists at IG Farben to make a slew of related compounds for him.[3]

The sulfa drugs were quickly recognized as active against a variety of streptococcal infections (Domagk's daughter was one of the first human patients treated). Domagk's 1935 paper on "A Contribution to the Chemotherapy of Bacterial Infections" became "not only a classic but—measured by strict experimental and statistical yardsticks—a masterpiece of careful and critical evaluation of a new therapeutic agent."[4] With sulfa drugs, the survival chances of patients afflicted with meningitis, certain pneumonias, and puerperal fever increased dramatically. In *The Youngest Science* Lewis Thomas describes the impact vividly in an autobiographical passage:[5]

> For most of the infectious diseases on the wards of the Boston City Hospital in 1937, there was nothing to be done beyond bed rest and good nursing care.
>
> Then came the explosive news of sulfanilamide, and the start of the real revolution in medicine.
>
> I remember the astonishment when the first cases of pneumococcal and streptococcal septicemia were treated in Boston in 1937. The phenomenon was almost beyond belief. Here were moribund patients, who would surely have died without treatment, improving in their appearance within a matter of hours of being given the medicine and feeling entirely well within the next day or so.
>
> The professionals most deeply affected by these extraordinary events were, I think, the interns. The older physicians were equally surprised, but took the news in stride. For an intern, it was the opening of a whole new world. We had been raised to be ready for one kind of profession, and we sensed that the profession itself had changed at the moment of our entry. We knew that other molecular variations of sulfanilamide were on their way from industry, and we heard about the possibility of penicillin and other antibiotics; we became convinced, overnight, that nothing lay beyond reach for the future.

(I will resist telling more about Gerhard Domagk except to mention that, like Boris Pasternak some decades later, Domagk was forced by a dictatorial regime to decline the Nobel Prize awarded to him in 1939.)

How do sulfa drugs work? By molecular mimicry. Folic acid or folate is an essential cell component in our bodies, a way station in the synthesis of more complicated molecules. We need it in our diet (it is in the B-vitamin family), but most bacteria make their own. They do so with enzymes that use p-aminobenzoic acid (illustration 10.3, left). Sulfanilamide is shown on the right; other sulfa drugs generate in the body molecules very much like it. Sulfanilamide resembles p-aminobenzoic acid, enough so as to fool the folic-acid-synthesizing enzymes of the bacteria, thus inhibiting their growth.

The essence of this tale of deception, found by chance for the sulfa drugs and now a strategic component of any drug-design scheme, is that the deception or inhibition need occur in a biological mechanism specific to the pathogen and not to the host.[6]

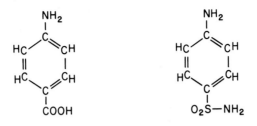

10.3 Shown above are p-aminobenzoic acid (*left*) and p-aminobenzenesulfonamide (sulfanilamide) (*right*).

The sulfa drugs were the first antibiotics. They preceded penicillin, but it could have been otherwise. To quote Erich Posner:

> Ironically enough, at that very time an agar plate containing an even more powerful antibacterial agent—penicillin—lay forgotten in St. Mary's Hospital in London. Its owner, Alexander Fleming, had become highly interested in prontosil and the sulfonamide derivatives that followed, but in his many papers on antibacterial and antiseptic treatments published between 1938 and 1940 he never mentioned penicillin, the antistaphylococcal action of which he had first observed in 1928. . . .
>
> Domagk was fortunate in having adequate chemical help, the lack of which prevented Fleming from advancing with penicillin for eleven years.[7]

4.

Where nerves meet striated muscle there is a neuromuscular junction, a gap. A signal needs to be transmitted; this is accomplished by small molecules, which diffuse merrily across the cleft between the nerve and muscle cells. Prominent among these molecules is acetylcholine, whose structure is shown in illustration 10.4:

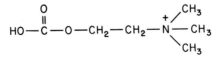

10.4 Acetylcholine, a neurotransmitter.

There are several "receptors" for acetylcholine in the muscle cell membrane. These are complex but are really nothing more mysterious than an assemblage of proteins or a channel in the cell membrane. Binding of acetylcholine to its receptor eventually (quickly) effects contraction of the muscle.

10.5 Curare being concocted by a Yanomamo Indian in the Amazonas Territory of Venezuela. Photograph courtesy of Robert W. Madden © National Geographic Society.

Curare is a venerable New World concoction of plant origin, used traditionally by South American Indians to poison their arrow tips. One of the active components (remember, everything is a mixture!) is d-tubocurarine, whose structure is shown in illustration 10.6:

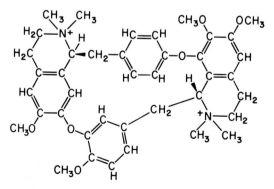

10.6 The molecular structure of d-tubocurarine.

This curare ingredient acts by competing effectively with acetylcholine for the latter's natural receptor sites. Once bound to the acetylcholine receptor, the curare ingredient fails to set into motion the chain of events that lead to muscle contraction. The nerve signal is in vain; the muscle is effectively paralyzed.

Occasionally, d-tubocurarine has been used as a muscle relaxant in surgery. A dose of only twenty to thirty milligrams causes a paralysis lasting for around thirty minutes. One has to be careful to provide artificial ventilation, for just as in the hunted prey of the South American Indians, it is the respiratory muscles that are paralyzed.[8]

Why does d-tubocurarine bind so effectively to a receptor that evolved to recognize acetylcholine? In the structure of the poison we gleam a hint: twice repeated in the ring is a nitrogen with a positive charge, bearing two CH_3's as well as another carbon (part of a ring). Such a "trialkylammonium" structural feature, uncommon in the biochemical world, is clearly perceived at one end of acetylcholine.

Jacob dressed in his brother's robes, with goat skins on his hands and neck, to make himself hairy. The feel and smell of Esau was enough to deceive Isaac, who was not a stupid or unperceptive man.

A piece of d-tubocurarine looks like a piece of acetylcholine. But my judgment of "looks like" is primitive—I spotted a single molecular grouping in these schematic pictures of a molecule. As already hinted in chapter 7 (and forming an important point to which I will return in a subsequent chapter), there is more than one way to represent (therefore recognize) a molecule. One could just draw lines between atomic symbols. Or try to sketch the three-dimensional shape of the molecule. Or represent, somehow, the bulk of its atoms, a so-called space-filling model. Or estimate the electric field emanating from it. Or one (well, another molecule, that is) could "stroke" the molecule. There are so many ways for us, or molecules, to "see" or "feel" one another. There are so many ways in which Isaac, or we, or a molecule, can decide whether "the other" is the same or not the same.

PART TWO

The Way It Is Told

11. The Chemical Article

Scientists have a nicely ambivalent attitude to the way their stories are told. On one hand, language is presumed immaterial—we supposedly have learned we need report the facts and nothing but the facts. Unambiguous mathematical equations and chemical structures make the story crystal clear, wherever on this globe it be told.

On the other hand, the language (whatever language we speak or write) is all we have. With it, written and spoken, we must convince the world that the knowledge we have gained with so much labor and ingenuity is indeed reliable—perhaps even superior to the knowledge of our ever-so-gentlemanly fellow workers in the field. Looking at the process of telling the story of chemical discovery or creation will reveal some important tensions underlying the molecular science.

You open an issue of a modern chemical periodical, say the important German *Angewandte Chemie* or the *Journal of the American Chemical Society*. What do you see? Riches upon riches: reports of new discoveries—marvelous molecules, unmakeable, unthinkable only yesterday, are made today and can be reproduced with ease. The chemist reads of

Chapters 11, 12, and 13 are adapted from Roald Hoffmann, "Under the Surface of the Chemical Article," in *Angewandte Chemie* 100 (1988): 1653–63, and *Angewandte Chemie (International Edition in English)* 27 (1988): 1593–1602.

the incredible properties of novel high-temperature superconductors, organic ferromagnets, and supercritical solvents. New techniques of measurement, quickly equipped with acronyms (e.g., EXAFS, INEPT, COCONOESY), allow you to puzzle out the structure of what you make more expeditiously. Information just *flows*. No matter if it's in German, or if it's in English. It's chemistry—communicated, exciting, alive.

11.1 The author (a little bit younger) reading chemical journals in Cornell University's Physical Sciences Library.

Let us take, however, another perspective. To the pages of these same journals turns a humanist, a perceptive, intelligent observer who has grappled with Shakespeare, Pushkin, Joyce, and Paul Celan. I have in mind a person who is interested in what is being written, and also in how and why it is written. My observer notes in the journal short articles only a page to ten pages in length. She notes an abundance of references, trappings familiar to literary scholars but perhaps in greater density (number of references per line text) than in scholarly texts in the humanities. She sees a large proportion of the printed page devoted to drawings. Often these seem to be pictures of molecules, yet they are curiously iconic, lacking complete atom designations. The chemist's representations are not isometric projections, nor real perspective drawings, yet they are partially three-dimensional.

My curious observer reads the text, perhaps defocusing from the jargon, perhaps penetrating it with the help of a chemist friend. She notes a ritual form: The first sentences often begin "The structure, bonding, and spectroscopy of molecules of type X have been subjects of intense interest.[a–z]" There is general use of the third person and a passive voice. She finds few overtly expressed personal motivations, and few accounts of historical development. Here and there in the neutered language she glimpses stated claims of achievement or priority— "a novel metabolite," "the first synthesis," "a general strategy," "parameter-free calculations." On studying many papers she finds a mind-deadening similarity. In the land of the new! Nevertheless, easy to spot in some of the articles, she also sees style—a distinctive, connected, scientific/written/graphic way of looking at the chemical universe.

Now, not hiding behind this observer, I want to take a look at the language of my science as it is expressed in the essential written record, the chemical journal article. I will argue that much more goes on in that article than one imagines at first sight; that what goes on is a kind of dialectical struggle between what a chemist imagines should be said (the paradigm, the normative) and what he or she must say to convince others of his argument or achievement. That struggle endows the most innocent-looking article with a lot of suppressed tension. To reveal that tension is (I will claim) not at all a sign of weakness or irrationality, but a recognition of the deep humanity of the creative act in science.

12. And How It Came to Be That Way

There was chemistry before the chemical journal. The new was described in books, in pamphlets or broadsides, in letters to secretaries of scientific societies. These societies—for instance, the Royal Society in London, chartered in 1662, or the Académie des Sciences, founded in Paris in 1666—played a critical role in the dissemination of scientific knowledge. Periodicals published by these societies helped to develop the particular combination of careful measurement and mathematization that shaped the successful new science of the time.[1]

The scientific articles of the period are a curious mixture of personal observation and discussion, with motivation, method, and history often given firsthand. Polemics abound. Cogent arguments for the beginning of a codification of the style of the scientific article in France and England in the seventeenth century have been given by Shapin, Dear, and Holmes.[2] I think the chemical article form rigidified finally in the 1830s and 1840s and that Germany was the scene of the hardening. The formative struggle was between the founders of modern German chemistry—people such as Justus von Liebig—and the Naturphilosophen. In that particular period the latter group might be represented by Goethe's followers, but their like was present elsewhere in Europe even earlier, in the eighteenth century. The "Nature-Philosophers" had well-formed notions, all-embracing theories, of how Nature should behave, but did not deign to get their hands dirty to find out what Nature actually did. Or they tried to fit Nature to their peculiar philosophical or poetic framework, not caring about what our senses and their extension (our instruments) said. The early nineteenth-century scientific article evolved to counter the pernicious influence of the Nature-Philosophers. The ideal report of scientific investigation should deal with the facts (often labeled explicitly or implicitly as truth). The facts had to be believable independent of the identity of the person presenting them. It followed that they should be presented unemotionally (thus, in the third person) and with no prejudgment of structure or causality (therefore the agentless or passive voice).

The fruits of this model reportage were immense. An emphasis on experimental facts stressed the reproducible. The conciseness of the German language seemed ideally suited for the developing paradigm. Cadres of chemists were trained. The development of the dyestuff industry that followed in England and Germany is a particularly well-studied manifestation of the industrial application of the new, organized chemistry.

The scientific article acquired in this period a canonical or ritual form. In illustration 12.1 I reproduce part of a typical article of that

12.1 An article by F. R. Goldman in *Berichte der Deutschen Chemischen Gesellschaft* 21 (1888): 1176–77.

period.[3] Note most of the features of a modern article—references, experimental part, discussion, diagrams. All that's lacking is the acknowledgment thanking the Deutsche Forschungsgemeinschaft or the National Science Foundation.

Through illustration 12.2, a contemporary article, we approach the present. This particular contribution, an important one by Wolfgang Oppolzer and Rumen Radinov, reports the synthesis of one, and specifically one, of the two enantiomers of muscone, a rare and valuable perfumery ingredient that normally is obtained from the male musk deer. The work is novel and significant, but I want to focus on the mode of presentation rather than the content.

J. Am. Chem. Soc. 1993, *115*, 1593–1594   1593

Scheme I

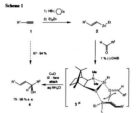

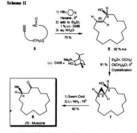

Synthesis of (R)-(−)-Muscone by an Asymmetrically Catalyzed Macrocyclization of an ω-Alkynal

Wolfgang Oppolzer* and Rumen N. Radinov

*Département de Chimie Organique, Université de Genève
CH-1211 Genève 4, Switzerland*
Received December 7, 1992

Recently, we reported a catalytic enantio-controlled approach to secondary (E)-allyl alcohols 4 (Scheme I).[1] Monohydroboration of alkynes 1 and "transmetalation" of the resulting (E)-(1-alkenyl)dicyclohexylboranes with diethylzinc conveniently provides (1-alkenyl)ethylzinc reagents 2. Nonisolated reagents 2 undergo exclusively π-face selective 1-alkenyl transfer to various aldehydes in the presence of 1 mol % of (−)-3-exo-(dimethylamino)isoborneol (DAIB).[2] This catalyst-directed 1-alkenyl/aldehyde addition is consistent with transition state 3*, where the zinc–aminoalkoxide chelate coordinates with both the aldehyde and the alkenylzinc reagent.[2] Were these two reactive units to be linked by a chain, one could expect a smooth macrocyclization to occur.

Hence, optically pure, macrocyclic (E)-allyl alcohols should be readily available from ω-alkynals in a single operation. This, however, requires that the dicyclohexylborane and the diethylzinc should react faster with the acetylene and alkenylborane functionalities, respectively, relative to their reaction with the aldehyde group.

We report here that this idea is feasible and applicable to the synthesis of enantiomerically pure (R)-muscone (8, Scheme II). This rare and valuable perfumery ingredient has been isolated from the male musk deer *Moschus moschiferus*,[3] and many syntheses of the racemate as well as of the natural antipode have appeared in the literature.[4]

Scheme II

75 %
91 %
82 %

(R)-Muscone

ω-Alkynal 5,[5] easily prepared by Swern oxidation[6] of 14-pentadecyn-1-ol[7] (95%), was added to a solution of freshly prepared dicyclohexylborane in hexane at 0 °C. After the reaction mixture was stirred at 0 °C for 2 h and diluted with hexane, the resulting 0.05 M solution of alkenylborane was added over 4 h to a 0.05 M solution of diethylzinc (1.5 mol equiv) in hexane containing (1S)-(+)-DAIB (0.01 mol equiv). Aqueous workup furnished the cyclic (S)-C$_{15}$-allyl alcohol 6[5] in 75% yield and in 92% ee.[8] To transfer the chirality from C(1) to C(3) we envisaged a hydroxy-directed cyclopropanation[9] using the Denmark protocol.[10] Indeed, slow addition of alcohol 6 to a mixture of Et$_2$Zn (2 mol equiv) and ClCH$_2$I (4 mol equiv) in 1,2-dichloroethane at 0 °C,

(1) Oppolzer, W.; Radinov, R. N. *Helv. Chim. Acta* 1992, *75*, 170.
(2) Reviews on asymmetric additions of organozinc reagents to aldehydes: Noyori, R.; Kitamura, M. *Angew. Chem.* 1991, *103*, 34; *Angew. Chem., Int. Ed. Engl.* 1991, *30*, 49. Soai, K.; Niwa, S. *Chem. Res.* 1992, *92*, 833.
(3) Isolation: Walbaum, H. *J. Prakt. Chem.* [2] 1906, *73*, 488. Structure: Ruzicka, L. *Helv. Chim. Acta* 1926, *9*, 715, 1008. Ställberg-Stenhagen, S. *Ark. Kemi* 1951, *3*, 517. Olfactive comparison of (R)- and (S)-muscone: Pickenhagen, W. *Flavor Chemistry, Trends and Developments*; ACS Symposium Series 388; American Chemical Society: Washington, DC, 1989; p 151.
(4) Syntheses of (R)-muscone: (a) Branca, Q.; Fischli, A. *Helv. Chim. Acta* 1977, *60*, 925. (b) Abad, A.; Arno, M.; Pardo, A.; Pedro, J. R.; Seoane, E. *Chem. Ind.* 1985, *29*. (c) Nelson, K. A.; Mash, E. A. *J. Org. Chem.* 1986, *51*, 2721. (d) Terunuma, D.; Motegi, M.; Tsuda, M.; Sawada, T.; Nozawa, H.; Nohira, H. *J. Org. Chem.* 1987, *52*, 1630. (e) Porter, N. A.; Lacher, B.; Chang, V. H.-T.; Magnin, D. R. *J. Am. Chem. Soc.* 1989, *111*, 8309. (f) Xie, Z.-F.; Sakai, K. *J. Org. Chem.* 1990, *55*, 820. (g) Tanaka, K.; Ushio, H.; Kawabata, Y.; Suzuki, H. *J. Chem. Soc., Perkin Trans. 1* 1991, 1445. (h) Ogawa, T.; Fang, C.-L.; Suemune, H.; Sakai, K. *J. Chem. Soc., Chem. Commun.* 1991, 1438. (i) Dowd, P.; Choi, S.-C. *Tetrahedron* 1992, *48*, 4773.

(5) All new compounds were characterized by IR, 1H NMR, 13C NMR, and mass spectra.
(6) Mancuso, A. J.; Huang, S.-L.; Swern, D. *J. Org. Chem.* 1978, *43*, 2480.
(7) Prepared by alkylation of bilithiated propargyl alcohol with 1-bromododecane (98%) and base-induced alkyne isomerization of resulting 2-pentadecyn-1-ol (93%): Utimoto, K.; Tanaka, M.; Kitai, M.; Nozaki, H. *Tetrahedron Lett.* 1978, 2301. Abrams, S. R. *Can. J. Chem.* 1984, *62*, 1333.
(8) Enantiomeric excess (ee) determined by 19F NMR of the (1S)-α-methoxy-α-(trifluoromethyl)phenylacetate: Dale, J. A.; Dull, D. L.; Mosher, H. S. *J. Org. Chem.* 1969, *34*, 2543. Absolute configuration of 6 assigned in analogy to the topicity of bimolecular 1-alkenylzinc/aldehyde additions[1] and confirmed by the conversion of 6 into (R)-muscone.
(9) Simmons–Smith reaction of (S)-cyclooctEN-3-ol: DePuy, C. H.; Marshall, J. L. *J. Org. Chem.* 1968, *33*, 3326. Diastereoselective cyclopropanations of cyclic (Z)-allyl alcohols: Poulter, C. D.; Friedrich, E. C.; Winstein, S. *J. Am. Chem. Soc.* 1969, *91*, 6892; 1970, *92*, 4274. NOESY correlation indicates a conformation of 6 with a syn-planar C(1)H/C(3)H orientation.
(10) Denmark, S. E.; Edwards, J. P. *J. Org. Chem.* 1991, *56*, 6974.

12.2 An article (first page of two) by W. Oppolzer and R. N. Radinov. Reprinted with permission from *J. Am. Chem. Soc.* 115 (1993): 1593. Copyright © 1993 American Chemical Society.

How does this article differ from one published a hundred years ago? The dominant language has changed, for interesting geopolitical reasons, to English. Yet it seems to me that there is not much change in the construction or tone of the chemical article. Oh, marvelous, totally new things continue to be reported. Measurements that took a lifetime are made in a millisecond. Molecules unthinkable a century ago are suddenly easily made, in a flash revealing to us their identity. And everything is now communicated with better graphics and computer typesetting, in a clearly flashier journal (though probably printed on poorer quality paper). But essentially the chemical article remains in the same form. Is that good, is that bad?

Well, I think both. The periodical article system of transmitting knowledge has worked remarkably well for two centuries or more. But there are real dangers implicit in its current canonical form. The article reports real facts, but at the same time it is unreal. It obscures the humanity of the process of creation and discovery in chemistry. Let me try to analyze what is "really" going in the writing and reading of a scientific article, which is much more than just communication of facts.

13. Beneath the Surface

On the face of it the article purports to be a communication of facts, perhaps a discussion, always dispassionate and rational, of alternative mechanisms or theories, and a more or less convincing choice between them. Or the demonstration of a new measurement technique, a new theory. And, remarkably, the article works. An experimental procedure detailed in a chemical journal in Russian or English can be reproduced (how easily it can be reproduced is another story) by someone with a rudimentary knowledge of either language working in Okazaki or Krasnoyarsk. This underlying feature of potential or real reproducibility is to me the ultimate proof that science is reliable knowledge.[1]

But in so many ways there is more than meets the eye in the scientific article. I see in it the following themes, many of which are also described and analyzed in a much deeper way in a remarkable book by David Locke, *Science as Writing*.[2]

The chemical article is a literary, therefore artistic, creation. Let me expand on what might be viewed as a radical exaggeration. What is art? Many things to many people. One aspect of art is aesthetic, another that it engenders an emotional response. In still another attempt to frame an elusive definition of that life-enhancing human activity, I will say that art is the seeking of the essence of some aspect of nature or of some emotion, by a human being. Art is constructed, human and patently unnatural. Art is intense, concentrated, in balance.

What is written in a scientific periodical is not a true and faithful representation (if such a thing were possible) of what actually transpired. It is not a laboratory notebook, and one knows that that notebook in turn is only a partially reliable guide to what took place. It is a more or less (one wishes more) carefully constructed, man- or woman-made *text*. Most of the obstacles that were in the way of the synthesis or the building of the spectrometer have been excised from the text. Those that remain serve the rhetorical purpose (no weaker just because it's suppressed) of making us think better of the author. The obstacles that are overcome highlight the success story.

The chemical article is a man-made, constructed abstraction of a chemical activity. If one is lucky it creates an emotional or aesthetic response in its readers.

Is there something to be ashamed of in acknowledging that our communications are not perfect mirrors but in substantial part literary texts? I don't think so. In fact, I think that there is something exquisitely beautiful about our texts. These "messages that abandon," to paraphrase Derrida,[3] indeed leave us, are flown to careful readers in every country in the world. There they are read, in their original language, and understood; there they give pleasure *and*, at the same time, they can be turned into chemical reactions, real new things. It would be incredible, were it not happening thousands of times each day.

One of the oft-cited distinguishing features of science, relative to the arts, is the more overt sense of chronology in science. It is made explicit in the copious use of references. But is it real history, or a prettified version?

A leading chemical style guide of my time admonished

> one approach which is to be avoided is narration of the whole chronology of work on a problem. The full story of a research may include an initial wrong guess, a false clue, a misinterpretation of directions, a fortuitous circumstance; such details possibly may have entertainment value in a talk on the research, but they are out of place in a formal paper. A paper should present, as directly as possible, the objective of the work, the results and the conclusions; the chance happenings along the way are of little consequence in the permanent record.[4]

I am in favor of conciseness, an economy of statement. But the advice of this style guide, if followed, leads to real crimes against the humanity of the scientist. In order to present a sanitized, paradigmatic account of a chemical study, one suppresses many of the truly creative acts. Among these are the human mind and hands responding to the "chance happening," the "fortuitous circumstance"—all of the elements of serendipity, of creative intuition at work.[5]

Taken in another way, the above prescription for good scientific style demonstrates very clearly that the chemical article is *not* a true representation of what transpired or was learned, but is a constructed text.

14. The Semiotics of Chemistry

Scientists think that what they say is not influenced by the national language (German, French, Chinese . . .) they use and the words within that language. The words employed, they think, are just representations of an underlying material reality which they, the scientists, have discovered or mathematicized. If the words are well chosen and precisely defined, they will be faithful representations of that reality, perfectly translatable, to any language.

That position *is* defensible: as soon as the synthesis of the new high-temperature superconductor $YBa_2Cu_3O_{~7}$ was described, it *was* reproduced, in a hundred laboratories around the globe.

But the real situation is more complex. The words we have, in any language, are ill-defined, ambiguous. A dictionary is a deeply circular device—just try and see how quickly a chain of definitions closes upon itself. Reasoning and argument, so essential to communication in science, proceed in words. The more contentious the argument, the simpler and more charged the words.

How does a chemist get out of this? Perhaps by realizing what some of our colleagues in linguistics and literary criticism learned over the last century.[1] The word is a sign, a piece of code. It signifies something, to be sure, but what it signifies must be decoded or interpreted by the reader. If two readers have different decoding mechanisms, then they will get different readings, different meanings. The reason that chemistry works around the world, so that BASF can build a plant in Germany or Brazil and expect it to work, is that chemists have in their education been taught the same set of signs.

I think this accounts in part for what C. F. von Weizsäcker noted in a perceptive article on "The Language of Physics":[2] If one examines a physics (read chemistry) research lecture in detail, one finds it to be full of imprecise statements, incomplete sentences, halts, and so on. The seminar is usually given extemporaneously, without notes, whereas humanists most often read a text verbatim. The language of physics or chemistry lectures is often imprecise. Yet chemists understand those presentations (well, at least some do). The reason is that the science lecturer invokes a code, a shared set of common knowledge. He or she doesn't have to complete a sentence—most everyone knows what he means halfway through that sentence.

The link between language and chemistry has always been there. So Lavoisier begins his revolutionary *Traité élémentaire de chimie* with a quote from the Abbé de Condillac: "We think only through the medium of words.—Languages are true analytical methods."[3] Lavoisier then reflects on his own work: "Thus, while I thought of myself employed only in forming a Nomenclature, . . . my work transformed itself by degrees, without my being able to prevent it, into a treatise upon the Elements of Chemistry."[4] The distinguished European writer Elias Canetti, author of a remarkable study of mass behavior (*Crowds and Power,* 1963) and a striking novel of the 1930s (*Auto-da-Fé,* 1935), earned a Ph.D. in chemistry. He credits chemistry with teaching him the importance of structure. And Benjamin Lee Whorf, the great American linguist, who made a case for language shaping culture, was an MIT-trained chemical engineer. Whorf was not averse to "an occasional chemical simile." In an essay on languages and logic, he writes: "the way the constituents are put together in these sentences of Shawnee and Nootka suggests a chemical compound, whereas their combination in English is more like a mechanical mixture."[5]

Pierre Laszlo has written a rich and original book explaining the language-chemistry link.[6] He posits an intriguing analogy between molecules and their transformations on one hand and linguistic structures such as morphemes, phonemes, ideograms and pictograms, transformations of mode, description, and so forth on the other. Laszlo's book

14.1   Portrait of M. and Mme. Lavoisier, by Jacques Louis David. © 1986.
The Metropolitan Museum of Art, Purchase, Mr. and Mrs. Charles Wrightsman
Gift, in honor of Everett Fahy, 1977. (1977.10), reproduced by permission.

goes further than a discussion of the uses of language in chemistry; it makes a plausible case for a parallel structure for chemistry and linguistics.

The semiotics of chemistry are most apparent in the structures of molecules that grace almost every page of a chemical journal, that identify, at a glance, a paper as chemical.[7] The given, for over a century, is that the structure of a molecule matters—not only the constituent atoms but how these atoms are connected up, how they are arranged in three-dimensional space, and how easily they move from their preferred equilibrium positions—and determines every physical, chemical, and ultimately biological property of the molecule.

It is crucial for chemists to communicate three-dimensional structural information to each other. The media for that communication are two-dimensional—a sheet of paper, a screen. So one immediately encounters the problem of representation.

15. What DOES That Molecule Look Like?

The structural information that chemists need to communicate is at some important level inherently graphic—it is essentially a shape to be drawn. And now we come to the crux of the matter. The group of professionals to whom this visual, three-dimensional information is essential are not necessarily talented (any more, any less than the average person) at transmitting such information. Chemists are not selected—and do not select themselves—for their profession on the basis of their artistic talents. Nor are they trained in basic art techniques. My ability to draw a face so that it looks like a face atrophied at age ten.

So how do they do it, how do we do it? With ease, almost without thinking, but with much more ambiguity than we, the chemists, think there is. The process is *representation,* a symbolic transformation of reality. It is both graphic and linguistic. It has a historicity. It is artistic *and* scientific. The representational process in chemistry is a shared code of this subculture.

We already saw one modern scientific paper (illustration 12.2). Let's look also at an informal drawing, the kind of information that passes between chemists when they talk, what is left behind in some restaurant, on a napkin or tablecloth, after dinner. Illustration 15.1 is such a drawing, by R. B. Woodward, a great organic chemist.

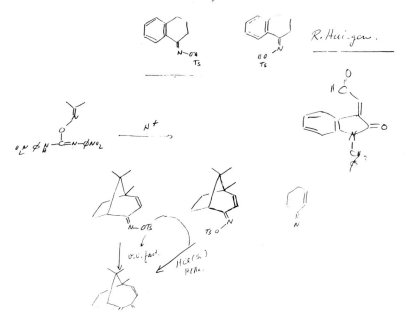

15.1 A drawing by R. B. Woodward, ca. 1966.

The substantial amount of graphic content just stares one in the face. There *are* little pictures here—lots of them—but the intelligent observer who is not a chemist is likely to be stymied. Indeed, one may find oneself in a situation analogous to that of Roland Barthes on his first visit to Japan, beautifully described in his *The Empire of Signs.*[1] What do these signs mean? We know that molecules are made of atoms, but what is one to make of a polygon such as in illustration 15.2, here repre-

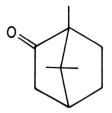

15.2 A typical chemical structure for camphor.

senting a white, waxy medicinal compound with a penetrating aroma (camphor)? Only one familiar atomic symbol, O for oxygen, emerges.

Well, it's a shorthand. Just as we get tired of saying the "United Na-

tions Educational, Scientific, and Cultural Organization" and write
UNESCO, so the chemist tires of writing all those carbons and hydrogens, ubiquitous elements that they are, and draws the carbon skeleton. Every vertex that is not specifically labeled otherwise in the structural representation above of camphor is carbon. Since the valence of
carbon (the number of bonds it forms) is typically four, chemists privy
to the code (and you) will know how many hydrogens to put at each
carbon. The polygon drawn in illustration 15.2 is in fact a graphic
shorthand for illustration 15.3.

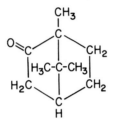

15.3 Camphor, with all the atoms specified.

But is illustration 15.3 the true structure of the molecule of camphor? Yes and no. At some level it is. At another level the chemist wants
to see the three-dimensional picture and so draws illustration 15.4.

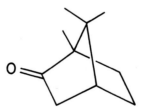

15.4 Camphor, a three-dimensional representation.

At still another level, he or she wants to see the "real" interatomic
distances (i.e., the molecule drawn in its correct proportions). Such
critical details are available, with a little money, and a little work, by the
technique I have already mentioned, called X-ray crystallography. And
so we have illustration 15.5, likely to be produced by a computer.[2]

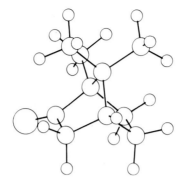

15.5   A ball-and-stick model of camphor.

This is a view of a so-called ball-and-stick model, perhaps the most familiar representation of a molecule in this century. The sizes of the balls representing the carbon, hydrogen, and oxygen atoms are some-what arbitrary. A more "realistic" representation of the volume that the atoms actually take up is given by the space-filling model of illustration 15.6. Note that in here the positions of the atoms (better said of their nuclei) become obscured. And neither illustration 15.5 or 15.6 is port-able—that is, it cannot be sketched by a chemist in the twenty seconds that a slide typically remains on a screen in the rapid-fire presentation of the new and intriguing by a visiting lecturer.

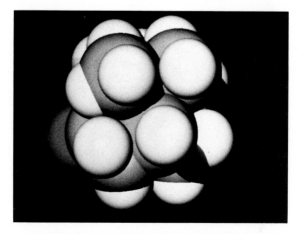

15.6   A space-filling model of camphor.

The ascending (descending?) ladder of complexity in representation hardly stops here. Along comes the physical chemist to remind her organic colleagues that the atoms are not nailed down in space, but are actually moving in near harmonic motion around those sites. The molecule vibrates; it does not have a static structure. Another chemist comes and says: "You've just drawn the positions of the nuclei. But chemistry is in the electrons. You should draw out the chance of finding them at a certain place in space at a certain time—the electronic distribution." This is attempted in illustration 15.7.

I could go on (the literature of chemistry certainly does). But let's

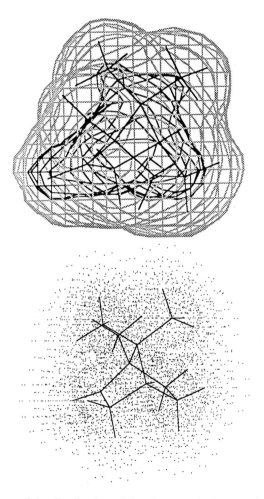

15.7 Two views of the distribution of the electrons in the camphor molecule.

stop and ask ourselves: which of these representations (you've seen seven!) is right? Which *is* the molecule? Well, all are, or none is. Or, to be serious—all of them are *models,* representations suitable for some purposes but not for others.[3] Sometimes just the name *camphor* will do. Sometimes the formula, $C_{10}H_{16}O$, suffices. Often it is the structure that's desired, and something similar to illustrations 15.2, 15.3, or 15.4 is fine. At other times one requires illustrations 15.5 or 15.6, or even either of the views in illustration 15.7.[4]

A final story needs to be told of camphor. This discussion comes from an article Pierre Laszlo and I wrote. We picked camphor as a molecule people might be familiar with but also one that had within it minimal complexities of representation. One of us (R.H.), having forgotten its structure, checked it in a textbook, then specified to some friends the geometry needed to produce the beautiful drawings of camphor you see. Every one of them was the mirror image of what you see, which *is* the naturally occurring material!

That we had the wrong "absolute configuration" was pointed out to us by a careful reader, Ryoji Noyori, the 1990 Baker Lecturer at Cornell. A literature search then revealed the wrong configuration disported by many, if not most, textbooks, including the Merck Index (a common chemical reference), and numerous literature papers. The structure is correctly given in the Sigma, Aldrich, and Fluka companies' chemical catalogues. Purveyors of goods are under more pressure to get their wares identified correctly than are their users!

16. Representation and Reality

Naive realism asserts that chemical formulas resemble reality: they do. It is possible to obtain pictures by physical means of benzene rings, one of the most common building blocks of organic molecules. They look, sometimes, like the benzene rings of the chemist (illustration 16.1). Sometimes they don't.

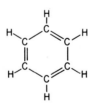

16.1 The structure of a benzene ring.

The scientist who thinks that now, with scanning tunneling microscopy (STM), a wonderful new tool, one can finally see atoms in molecules, has a shock coming when he or she looks at a scanning tunneling microscope image of graphite.[1] Graphite is built up of two-dimensional "chicken-wire" nets made up of benzene rings. You should see six carbons. But half the hexagonal lattice atoms are highlighted in the STM image, half not—and for good reasons.[2] Seeing and believing have a complex relationship to each other. The benzene rings of the chemist are rough approximations. They stand not unlike a metaphor to the molecular object represented.

Let us fix on the *typical* level of presentation (of illustrations 15.2 or 15.4), that of a polygon or a three-dimensional idealization of it. What *are* these curious drawings, filling the pages of a scientific paper? I now ask the question from the point of view of an artist or draftsman. They are not isometric projections, certainly not photographs. Yet they're obviously attempts to represent in two dimensions a three-dimensional object for the purpose of communicating its essence to some remote reader.

It is fascinating to see the chemical structures on the pages of every journal and to realize that from such minimal information people can actually *see* molecules in their mind's eye. The clues to three-dimensionality are minimal. The molecules float (illustration 16.2, left; the molecule illustrated is norbornane, C_7H_{12}, the heart of camphor), and you're usually discouraged from putting in a reference set of planes to help you see them (center).

16.2   Three views of norbornane, C_7H_{12}.

Some chemists rely so much on the code that they don't draw norbornane as on the left side of illustration 16.2, but as on the right. What's the difference? One line "crossed" instead of "broken." What a trivial clue to three-dimensional reconstruction, that one part of a molecule is put behind another one! This is hardly a modern invention, something that one must learn at the Ecole des Beaux Arts. Illustration 16.3 shows one of the cave paintings from Lascaux.[3] Note the treatment of the legs of the bisons as they go into the body. Now surely smart chemists should be able to do what cavemen 15,000 years ago did, shouldn't they? Too often they don't bother.

16.3 Cave painting of two bison, Lascaux. © Artists Rights Society (ARS), C.N.M.H.S./S.P.A.D.E.M.

Scattered about in the drawings of chemists (e.g., illustration 12.2) are sundry wedges and dashed lines. As noted in chapter 9, these are pieces of a visual code, simple in conception: a solid line is in the plane of the paper, a wedge extends out in front, a dashed line extends to the back. Thus in chapter 9, illustration 9.1 showed one view, quite recognizable to chemists, of the tetrahedral methane molecule, CH_4. The tetrahedron is the single most important geometrical figure in chemistry.

Describing this notation may be enough to make these structures rise from the page for some people, but the neural networks that control representation are effectively etched in, for life, when one handles (in human hands, not in a computer) a ball-and-stick model of the molecule while looking at its picture.

A glance at the more complicated molecules of illustration 12.2 shows that the wedge-dash convention is not applied consistently. Most compounds have more than a single plane of interest; what's behind one plane may be in front of another. So the convention is almost immediately used unsystematically, the author or lecturer choosing to emphasize the plane he or she thinks important. The result is a cubist perspective, a kind of Hockney photo-collage where the same thing is shown from several different perspectives.[4] The molecule is certainly seen, but it may not be seen as the chemist thinks (in a dogmatic moment) that it is seen. It is represented as he chooses to see it, nicely superimposing a human illogic on top of an equally human logic.

The policies of journals, their economic limitations, and the available technology put constraints not only on what is printed but also on how we *think* about molecules. Take norbornane (illustration 16.4, right). Until about 1950 no journal in the world was prepared to reproduce this structure as shown in illustration 16.4 (right). Instead you saw it in the journal as illustration 16.4 (left). Now everyone had known since 1874 that carbon is tetrahedral, meaning that the four bonds to it are formed along the four directions radiating out from the center of a tetrahedron to its vertices. Molecular models were available or could be relatively easily built. Yet I suspect that the icon of norbornane that a typical chemist had in his mind around 1925 was that on the left of illustration 16.4, not that on the right. He was conditioned by what he saw in a journal or textbook—an image, and a flat one at that. He might have been—I think he often was—moved to act (in synthesizing a derivative of this molecule, for instance) by that unrealistic two-dimensional image.

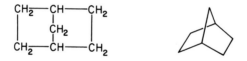

16.4 Norbornane.

Maybe it's not that different from the way we approach romance in our lives, equipped with a piecewise reliable set of images from novels and movies.

17. Struggles

Forces contend, perforce, under the surface of the chemical article. This is unavoidable, for science depends on argument. *Argument* has several meanings: it could be taken as a simple process of reasoning, a statement of fact; or the word also may mean disagreement, the confrontation of opposites. I would claim that both senses are essential to science: dispassionate logical reasoning *and* impassioned conviction that one (model, theory, measurement) is right and another is wrong. I feel that scientific creativity is rooted in the inner tension, within one and the same person, of knowing that he or she is right and knowing that that conviction has to be proven to the satisfaction of others—in a journal article.

A nice, even-toned, scientific article may hide strong emotional undercurrents, rhetorical maneuvering, and claims of power. The desire to convince by screaming, "I'm right, all of you are wrong" clashes with the established rules of civility supposedly governing scholarly behavior. Where this balance is struck depends on the individual.

Another unvoiced dialogue exists between experiment and theory. There is nothing special about the love-hate relationship between experimentalists and theorists in chemistry. You can substitute "writer" and "critic" and talk about literature, or find the analogous characteris-

17.1 Cartoon by Constance Heller.[1]

tics in economics. The lines of the relationship are easily caricatured: Experimentalists think theorists are unrealistic and build castles in the sky; yet they need the frameworks of understanding that theorists provide. Theorists may distrust experiments and wish people would do that crucial measurement the theorist wants. But where would the theorists be without any contact with reality?

17.2 Cartoon by Constance Heller.

An amusing manifestation of the feelings about the theory-experiment waltz is to be found in the occasionally extended quasi-theoretical discussion sections of experimental papers. These sections in part contain a true search for understanding, but in part what goes on in them is an attempt to use the accepted reductionist ideal (with its exaggerated hailing of the more mathematical) so as—not to put too fine a point on it—to impress one's colleagues. On the other side, I often put more references to experimental work in my theoretical papers than I should, because I'm trying to "buy credibility" from my experimental audience. If I show experimental chemists that I know of their work, perhaps they will listen to my wild speculations.

Another related struggle is between pure and applied chemistry. It's interesting to reflect that this separation may have also had its roots in Germany in the mid-nineteenth century; it seems to me that in the other chemical power of that time, Britain, the distinction was less congealed. Reaching out after some justification in terms of industrial use is quite typical in a pure chemical paper. But at the same time, there is a falling back, an unwillingness to deal with the often unruly, tremendously complicated world of, say, industrial catalysis. And in industrial settings there is a reaching after academic credentials.

Perhaps the greatest unvoiced struggle in chemistry, a science closely linked to the economy, is whether to reveal or to conceal. This is not a struggle visible as such in the chemical literature, for once the decision to publish has been made, what is published had better be right. The more interesting your result, the more likely your competitors are to verify it. And you can be sure they'll publicize your mistakes with glee.

No, the critical decision, should you have something of commercial value, is whether to delay publishing to "establish your patent position," or perhaps not publish at all. Recall the remarkable story of Gerhard Domagk's discovery of the sulfa drugs, told in chapter 10. Domagk was working for a German conglomerate, IG Farbenindustrie. His extraordinary 1935 paper on the first such drug, Prontosil, reports experiments done three years before. Domagk published one month after the patent was issued, not before.[2]

18. The Id Will Out

I use *id* here in the psychoanalytic sense, referring to the complex of instinctive desires and terrors that inhabit the collective unconscious. On the one hand these irrational impulses—aggression figuring most prominently—are our dark side. On the other hand, they provide the motive force for creative activity.

Science is done by human beings and their tools. Which means that it is done by fallible human beings. The driving forces for acquiring knowledge are, to be sure, curiosity and altruism, rational motives. But creation is just as surely rooted in the irrational, in the dark, murky waters of the psyche where fears, power, sex, and childhood traumas swim in all their hidden, mysterious movements. And spur us on. Not only do character and deep-down motivations matter—their "unsavory" side may well be the driving force of the creative act. Mind you, this is not a justification for being unethical; to be a good human being is as important an aspiration for a scientist as for anyone. But scientists are no better than anyone else, just because they're scientists.

The irrational seems to be effectively suppressed in the written sci-

This chapter is adapted from Roald Hoffmann, "Under the Surface of the Chemical Article," *Angewandte Chemie* 100 (1988): 1653–63, and *Angewandte Chemie (International Edition in English)* 27 (1988): 1593–1602.

entific word. But of course scientists are human, no matter how much they might pretend in their articles that they are not. Their inner illogical forces push out. Where? If you don't allow them in the light of day, on the printed page, then they will creep out or explode in the night, where things are hidden, and no one can see how nasty you are. I refer, of course, to the anonymous "refereeing" process. When I submit a paper to a chemical journal, the editor sends it out to at least two reviewers, presumably experts in my field. In due time (much, much more quickly, incidentally, than in literary or humanities journals, with which I have some experience), I will receive the anonymous comments of the reviewers.

In the course of this refereeing process there are incredibly irrational responses unleashed by perfectly good and otherwise rational scientists. Here is a selection of some I've received:

Paper 1: "The speculations in this paper are the sort of thing that one expects to hear at research seminars, or in social chemical gatherings over a glass of beer; certainly many of them have been made at my own seminar by bright young students. No one else, however, has had the conceit or effrontery to think them worth publishing, let alone in a communication written in the first person. This paper seems to me entirely unsuitable for publication in any reputable scientific journal, let alone JACS [*Journal of the American Chemical Society*]."

Paper 2 (a paper submitted to a chemical journal but reviewed by a physicist): "This paper would not be acceptable for publication in *Physical Review*. The authors should calculate the binding energy of this structure and compare it with graphite, not just propose it as a possible structure. The extended Hückel method contains errors of order 3eV; it is absolutely useless except for publishing papers in chemistry journals. You chemists should raise your standards."

Paper 3: "I am not now and never have been an admirer of Hoffmann's efforts in the inorganic/organometallic field. To a bridge player, the sideline kibitzers, however intelligent, are of little interest. Hoffmann is very intelligent—but not intelligent enough to do anything positive. Kibitzing, however well done, soon becomes tiresome.

I also wonder why he has such a tremendous ego as to suppose that *everything* he does belongs in JACS. This paper, to take a case in point, does *not* belong there. It is just the nth in a long, very familiar line."

Three hundred papers down the line, I can take it pretty well. But in the beginning these were devastating. Of course, my own reviewer's comments are completely rational and gentlemanly (I'm smiling).

Actually, most of the comments I get on my papers are not as bereft of substance as these. When,they make me angry I try to think of the contrast with the reviewer's comments that come with my poetry rejection slips. Usually there are *no* such comments, none at all. Just a rejection.

I actually think that what saves the chemical article from complete dullness is that its language comes under stress. We are trying to communicate things in words that perhaps cannot be expressed in words but require other signs—structures, equations, graphs. And we are trying hard to eliminate emotion from what we say—which is impossible. So the words we use occasionally become supercharged with the tension of everything that's *not* being said.

So, many things go on in a scientific article, on and under the surface. Now let me turn from the way things are said (and the underlying tensions revealed in writing and representing structure in chemistry) and turn to the chemistry reported. It's novel, but is it discovery?

PART THREE

Making Molecules

19. CREATION AND DISCOVERY

In describing what they do, scientists have by and large bought the metaphor of discovery, and artists that of creation. The cliché "uncovering the secrets of nature" has set, like good cement, in our minds. But I think that the metaphor of discovery is effective in describing only part of the activity of scientists, a smaller piece still of the work of chemists. There are historical, psychological, and sociological reasons for the ready acceptance of the metaphor, and these need to be brought out.

History and psychology: The rise of modern science in Europe coincided with the age of geographical exploration. Men set foot on distant shores, explored terra incognita. Even in our century, the man I was named after first sailed the Northwest Passage and reached the South Pole. Voyages of discovery, maps filled in, those are powerful images indeed. So is the first gaze into a royal tomb full of glistening gold vessels. It is no surprise that these metaphors were and are accepted by scientists as appropriate descriptors of their generally laboratory-bound activity. Is there some vicarious sharing of imagined adventures at work here?

This chapter was adapted from one first published in *American Scientist* 78 (1990): 14–15.

Here is a typical expression of the attitude of the time, by a great chemist who also wrote poetry, Humphry Davy:

> Oh, most magnificent and noble Nature!
> Have I not worshipped thee with such a love
> As never mortal man before displayed?
> Adored thee in thy majesty of visible creation,
> And searched into thy hidden and mysterious ways
> As Poet, as Philosopher, as Sage?[1]

The male metaphors of peeking, unveiling, penetrating are a characteristic of nineteenth-century science. They fit the idea of discovery.

Sociology and Education: Those philosophers of science who started out as practicing scientists have generally, I believe, come from physics and mathematics. (There is one exception; Michael Polanyi, a distinguished philosopher, was an insightful physical chemist.) The education of professional philosophers is likely to favor the same fields; there is a special role, quite understandably, for logic in philosophy. No wonder that the prevailing ideology of reasoning in the underlying scientific areas of expertise of philosophers of science has been extended by them, unrealistically I believe, to all science.

Philosophy: The French rationalist tradition, and the systematization of astronomy and physics before the other sciences, have left science with a reductionist philosophy at its core. This I've already fought with before, in chapter 4. The logic of a reductionist philosophy fits the discovery metaphor—one digs deeper and discovers the truth.

But reductionism is only one face of understanding. We have been made not only to disassemble, disconnect, and analyze but also to build. There is no more stringent test of passive understanding than active creation. Perhaps "test" is not the word here, for building or creation differ inherently from reductionist analysis. I want to claim a greater role in science for the forward, constructive mode.[2] Richard Feynman once wrote on his blackboard "What I cannot create I do not understand."[3]

And Goethe, in his unique 1809 novel *Elective Affinities,* constructed on the metaphor of a theory of chemical bonding, in a prescient tribute to synthesis when analysis was still at the center of the science, creates this conversation between Eduard and Charlotte:

> "[T]he affinities become interesting only when they bring about divorces."

"Does that doleful word, which one unhappily hears so often in society these days, also occur in natural science?"

"To be sure," Eduard replied. "It even used to be a title of honour to chemists to call them artists in divorcing one thing from another."

"Then it is not so any longer," Charlotte said, "and a very good thing too. Uniting is a greater art and a greater merit. An artist in unification in any subject would be welcomed the world over."[4]

19.1 The principal characters in Goethe's "Elective Affinities."
Original drawing by H. A. Dähling, engraved (1811) by Heinrich Schmidt.

What is strange is that chemists should accept the metaphor of discovery. Chemistry is the science of molecules (up to a hundred years ago one would have said "substances" or "compounds") and their transformations. Some of the molecules are indeed *there,* just waiting to be known by us. "Known" in their static properties—what atoms are in them, how these are connected up, the shapes of molecules, their splendid colors. And in their dynamic characteristics: the molecules' internal motions, their reactivity. The molecules are those of the earth—for instance, simple water and complex malachite. Or of life—pretty simple cholesterol, more complicated hemoglobin. The discovery paradigm certainly applies to the study of these molecules.

But so many more molecules of chemistry are made by us, in the laboratory. We're awfully prolific—a registry of known, well-characterized compounds now numbers over ten million. Ten million compounds that were not on earth before! It is true that their constitution follows underlying rules, and if chemist A had not made such-and-such a molecule on a certain day, then it is likely to have been synthesized a few days or decades later by chemist B. But the human being, a chemist, chooses the molecule to be made and a distinct way to make it.[5] The situation is not all that different from the artist who, constrained by the physics of pigment and canvas, and shaped by his or her training, nevertheless creates the new.

Even when one is operating clearly in the discovery mode in chemistry, elucidating the structure or dynamics of a known, naturally occurring molecule, one usually has to intervene with created molecules. I once heard a beautiful lecture by Alan Battersby, an outstanding British organic chemist, dealing with the biosynthesis of uroporphyrinogen-III (even in the trade, the name of this molecule is abbreviated as uro'gen-III). It isn't a glamorous molecule, but it should be. For from this precursor plants make chlorophyll, the basis of all photosynthetic activity. All cells use another uro'gen-III derivative in cytochromes for electron transport. And the crucial iron-containing oxygen carrier piece of hemoglobin derives from this small disk-shaped molecule.

Uro'gen-III, pictured in illustration 19.2, is made from four rings, called pyrroles, themselves tied into a larger ring. Note the markers A and P in each ring. The letters stand for *A*cetyl (CH_2COOH) and *P*ropionyl (CH_2CH_2COOH) atom groupings. They are in the same order as one goes around the ring (from around ten o'clock)—except for the last set, which are "reversed." So the markers read A—P, A—P, A—P, P—A.

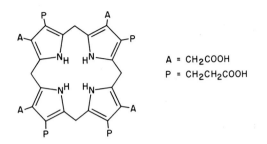

$A = CH_2COOH$

$P = CH_2CH_2COOH$

19.2 Uroporphyrinogen-III (uro'gen III).

How this natural molecule is assembled, within us, is clearly a discovery question. In fact, the four pyrrole rings are connected up, with the aid of an enzyme, into a chain, then cyclized. But the last ring is first put in "incorrectly" (i.e., with the order of the A,P labels the same as in the other rings, A—P, A—P, A—P, A—P). Then, in a fantastic separate reaction sequence, just that last ring, with its attached labels, is flipped into position.

This incredible but true story was deduced by Battersby and coworkers using a sequence of synthetic molecules, not natural ones.[6] Each was designed to model some critical way station molecule in the living system. Each synthetic hypothetical intermediate was subjected to conditions resembling physiological ones, eventually tracing out the sequence of the natural processes. Thus, using molecules—unnatural ones—that *we* have made, we've learned how nature builds a molecule that makes life possible.

The synthesis of molecules puts chemistry very close to the arts. We create the objects that we or others then study or appreciate, a thought voiced over a hundred years ago by Marcellin Berthelot.[7] That's exactly what writers, composers, and visual artists, working within their areas, do. I believe that, in fact, this creative capacity is exceptionally strong in chemistry. Mathematicians also study the objects of their own creation, but those objects, not to take anything away from their uniqueness, are mental concepts rather than real structures. Some branches of engineering are actually close to chemistry in this matter of synthesis. Perhaps this is a factor in the kinship the chemist-narrator feels for the builder Faussone, who is the main character of Primo Levi's novel *The Monkey's Wrench*.[8]

The distinctive constructive nature of engineering is clear in the following analysis by David Billington:

> Science and engineering may share the same techniques of discovery—physical experiments, mathematical formulation—but students quickly learn that the techniques have vastly different applications in the two disciplines. Engineering analysis is a matter of observing and testing the actual working of bridges, automobiles, and other objects made by people, while scientific analysis relies on closely controlled laboratory experiments or observations of natural phenomena and on general mathematical theories that explain them. The engineer studies objects in order to change them; the scientist, to explain them.[9]

Too bad that Billington falls into the familiar trap of typing science as discovery.

The building of theories and hypotheses is a creative act as well, even more so than synthesis. One has to imagine, to conjure up a model that fits often irregular observations.[10] There are rules; the model should be consistent with previously received reliable knowledge. There are hints of what to do; one sees what was done in related problems. But what one seeks is an explanation that was not there before, a connection between two worlds. Often, actually, it is a metaphor that provides the clue: "Two interacting systems, hmm . . , let's model them with a resonating pair of harmonic oscillators, or . . . a barrier penetration problem."[11] The world out there is moderately chaotic, frighteningly so in the parts we do not understand. We want to see a pattern in it. We're clever, we "connoisseurs of chaos,"[12] so we find/create one. A sensitive reader, Mary Reppy, made an astute comment here:

> I think that there is a balance between admitting enough of the complexity of reality into a problem for it to be interesting, while keeping the chunk of reality one is considering simple enough (through approximations) or small enough so that it *can* be modeled. A completely understood (or "reduced") problem is boring, but a realistically complex one is frustrating.[13]

Had more philosophers of science been trained in chemistry, I'm sure we would have a very different paradigm of science before us.

Is art all creation? I don't think so. I note here the work of Eliseo Vivas, who has written a book of essays with the same title as the present chapter. Vivas argues that much of art is a merged process of discovery in creation. In an essay on poetry he says that the poet does *not* do what the writer of Genesis reports God did when he tells us that "In the beginning God created the heaven and the earth." Rather what the poet does is more like what he (God) is reported to have done in the second verse. Before the poet comes along the earth, for us, is without form and void, and darkness is upon the face of the deep. The poet divides the light from the darkness and gives us an ordered world. If it were not for him we would not see it. . . . The poem reveals to us what the poet discerns through an act of creation.[14]

And he continues:

I conceive of the creative mind as discovering subsisting values. . . . From the standpoint of culture, the mind creates new values, for these were not there before for the creative mind or the culture. But the mind discovers them by bringing them up from the realm of subsistence, into the poem, from where they get carried away by its readers and put into a circulation, so to speak, into the market place.[15]

And Richard Moore, a poet, writes:

The artist had best say every prayer to every unknown force and perform every ritual in the hope that he or she will create nothing, only find for all who are interested what is there to find. He must not create, but discover.[16]

I agree with Vivas and Moore. I think art is in substantial measure discovery—of the deep truths of what is also around us, often overlapping, but more often reaching outside the set of problems that science has set for itself to try to understand. Art aspires to discover, explore, unravel—whatever metaphor you please—the nonunique, chanced, irreducible world within us: "To build ever more perfect lives, invisible cities, our own constellations."[17]

20. In Praise of Synthesis

Creation is wonderful. We admire Nature's work first—from simple things such as the hoarfrost that settled overnight on the red of these blooming maples, to that most intricate creation, repeated thousands of times each day, a human infant brought to term and born. We admire human creation second—Mozart and his librettist Lorenzo da Ponte, soprano Elly Ameling and an English orchestra, two hundred years apart, collaborate on a rendition of "Voi che sapete" that is so sweet and clear that it almost hurts. Or David Hockney, assembling some fifty roughly developed prints into a photo-collage in which the camera, Hockney, and we, like the eye, concentrate on a detail here, jump there, zoom in on a piece of the background. Or Phil Eaton and Thomas Cole, who synthesized a simple molecule, cubane, which has eight carbon atoms in the shape of a cube, each carbon also bearing a hydrogen (see illustration 20.1).

This chapter is adapted from one of the same title first published in *Negative Capability* 10, nos. 2–3 (1990): 162–75.

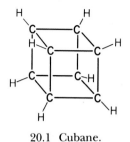

20.1 Cubane.

I want to praise chemical synthesis, the making of molecules. Synthesis is a remarkable activity that is at the heart of chemistry, that puts chemistry close to art, and yet has so much logic in it that people have tried to teach computers to design the strategy for making molecules.

Chemists make molecules. They do other things, to be sure—they study the properties of these molecules; they analyze as we have seen, they form theories as to why molecules are stable, why they have the shapes or colors that they do; they study mechanisms, trying to find out how molecules react. But at the heart of their science is the molecule that is made, either by a natural process or by a human being.[1]

There is not one way to make molecules, but many. So let's look at some different kinds of chemical synthesis. These are shaped by scientific needs, economic considerations, traditions, and aesthetics.

1. Elemental. You take substance A, perhaps an element, perhaps a compound, mix it with substance B, beat on it with heat, light, zap it with an electrical discharge. In a puff of foul smoke, a flash, an explosion, out pop lovely crystals of desired substance C. This is the comic-book stereotype of chemical synthesis (illustration 20.2). In general, an elemental synthesis is not considered by the chemical community to be a clever way of making molecules. Unless, unless—the product molecule had not been on earth before. This is how XeF_4 was made, with no pyrotechnics, but still by an elemental synthesis:[2]

$$Xe + 2F_2 \xrightarrow{\text{heat}} XeF_4$$

20.2 A well-known chemist at work. From "Walt Disney's Donald Duck Adventures," story and drawings by Carl Barks, colored by Sue Daigle, no. 15 (September) (Prescott, Ariz.: Gladstone, 1989). This is a reprint of issue 44 of *Walt Disney's Comics and Stories*, first published in 1944, © The Walt Disney Company.

Behind its creation was some clever reasoning by Neil Bartlett,[3] which allowed the makers of XeF_4 to imagine that the compound might exist. It was the first simple compound of a noble gas and a halogen. The uniqueness of creation can override stylistic reservations as to how the product is made.

2. Part by design, part by chance. In that limbo between serendipity and logic, there stirs the vast majority of chemical syntheses. One has a rough idea of what one wants to do—cleave a bond there, form a new one here. One has read of similar reactions on molecules that look vaguely like the one at hand, and so one tries (or more likely asks a graduate student to try) one of those reactions. It might work, it might not—perhaps the conditions must be juggled, the temperature changed, or one should follow a different regime of adding the reagents, to give them more or less time to mix. On the seventh runthrough, something happens. There is mostly insoluble brown gunk in the reaction vessel, but if one separates the liquid, extracts it with another solvent, allows the material to crystallize, out come translucent lilac crystals of a product.

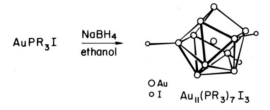

$$AuPR_3I \xrightarrow[\text{ethanol}]{NaBH_4}$$

O Au
o I $Au_{11}(PR_3)_7 I_3$

20.3 Synthesis of a gold cluster. Each "outside" gold atom carries a PR_3 group radiating out of the center of the cluster. These are not shown, so as to reveal the central cluster core.

An example of such a synthesis is a reaction (see illustration 20.3) in which a spectacular gold cluster formed. The chemists in Milan who did this[4] began with a simple gold phosphine iodide. They subjected it to reaction conditions ($NaBH_4$, ethanol) that in some other cases had led to novel gold-gold bonding. The synthesizers had an idea that something interesting might happen. But I think it is fair to say that they did not anticipate exactly what *did* happen, even though—and this is very important—they were well prepared to follow up and determine just what molecules were created in their flask. In fact, a marvelous cluster, one gold atom in the middle, with ten other gold atoms (an icosahedron minus two) on the outside, assembled itself.

3. Industrial synthesis. Illustration 20.4 shows one way aspirin is made commercially. The number of pills manufactured in the United States per year approaches the number of dollars in the defense budget

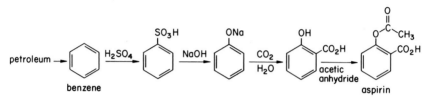

20.4 One commercial synthesis of aspirin.

From a petroleum fraction, benzene is separated out, then reacted sequentially with sulfuric acid, sodium hydroxide (lye), dry ice and water, and acetic anhydride (vinegar hiding) to yield acetylsalicylic acid, which is aspirin.

Some years ago, *Punch* made an apt verse commentary on synthesis and what are called "chemical feedstocks":

> There's hardly a thing a man can name
> Of beauty or use in life's small game,
> But you can extract in alembro or jar,
> From the physical basis of black coal tar:
> Oil and ointment, and wax and wine,
> And the lovely colours called aniline:
> You can make anything, from salve to a star
> (If only you know how), from black coal tar.

The making of aspirin, like most fine chemical manufacture, begins with a portion of refined petroleum. Right there is a problem, and a challenge—how to make those complex structures from sources less easily depleted.

An important factor in any industrial synthesis is safety. The manufacturing process must not injure the health of the workers, nor, as we

20.5 An oil refinery, photo by Robert Smith. Many of the chemicals used in synthesis are derived from petroleum. (Tony Stone Images)

have realized only slowly, the environment; the final product must be safe to the consumer. In this context, people have speculated whether aspirin would be allowed as a nonprescription medication were it put on the market today.

The overriding imperative in industrial synthesis is cost. Starting products and reagents had better be as close as possible to earth, air, fire, and water (and fire is getting awfully expensive). All the reagents in the aspirin synthesis are on the "top fifty" list of the chemical production hits chart—in volume of production and in least cost. Expense also drives producers to optimize the efficiency of synthesis. If a step in a synthesis gives a yield of 90 percent (that is, 90 percent of the theoretically possible amount: more on yields in the next chapter), then an improvement to 95 percent, through a new catalyst, might provide a competitive advantage of millions of dollars. In the past, this led to strategies such as "take the next chemical off the shelf and try it." Today the progressive segment of industry invests in basic studies of the way chemical reactions proceed, the rational path to improve a process.

The competitive pressure to reduce cost is also the source of much creativity in industrial synthesis. The academic chemist can and will flit to the next exciting problem if one synthesis doesn't work out. The industrial chemist doesn't have that choice. So he or she pushes on, often to ingenious solutions.[5]

21. Cubane, and the Art of Making It

There is another kind of synthesis that is planned, just as industrial syntheses are. Many of the masterpieces of synthesis are created in an academic setting. The constraints of cost are relaxed, though they are still there. Imagination is set free. Marvelous syntheses result. Here is one (already mentioned), that of cubane. This carbon die is an unnatural product; it was made not because it was thought useful, but because it is beautiful, in a simple Platonic solid sense. It was also made because it was there, waiting to be made, like the proverbial mountain. Others failed to synthesize it before two people at the University of Chicago succeeded in 1964.[1]

Here is the chart (illustration 21.1) from the original paper showing how Eaton and Cole did it. We have before us ten molecules with nine arrows, or reactions, between them. Above each arrow is the briefest mnemonic description of the reaction conditions. Each reaction might be composed of five to twenty distinct physical manipulations: weighing out reagents; dissolving them in a solvent; mixing, stirring, and heating; filtration; dessication; and so on. A step might take an hour or a week. And the scheme does not include the laborious and ingenious analytical chemistry required to identify those intermediate molecules.

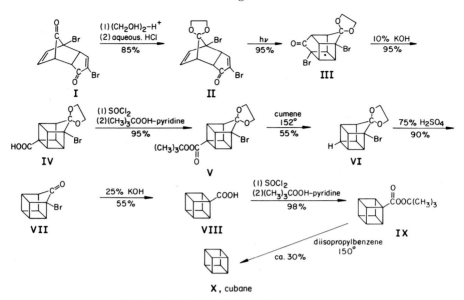

21.1 The Cole and Eaton cubane synthesis.

At the end of the synthesis is cubane. At the beginning of the synthesis is molecule I. It doesn't look simple—one thinks that at the start of any construction there should be readily available materials. Actually, starting material I *is* easy to make. The Chicago pair synthesized it earlier, in three steps, from another molecule that costs pennies per gram.

Below each arrow is a percentage. This is the yield, the percent of the theoretically possible product that is actually obtained.

Supposing you run the following reaction: a molecule C_4H_6 (butadiene, it happens to be called) is transformed to C_6H_{10} (cyclohexene) by the addition of a molecule, C_2H_4 (ethylene), with two carbons and four hydrogens.

$$C_4H_6 + C_2H_4 \rightarrow C_6H_{10}$$

If on some scale (the atomic mass scale) a hydrogen weighs one unit and a carbon atom twelve times as much, as they do, you would go from $4 \cdot 12 + 6 \cdot 1 = 54$ atomic mass units for C_4H_6 to $6 \cdot 12 + 10 \cdot 1 = 82$ atomic mass units for the product C_6H_{10}. The actual weights would depend on the number of C_4H_6 molecules you began with—you could have one gram of C_4H_6, one ton, or one milligram. Whatever mass of C_4H_6 is in your pot in the beginning, the *most* C_6H_{10} you could get out of C_4H_6 in the end is 82/54 of that original mass. There is no way to make something out of nothing. Matter is conserved; there are no nuclear reactions here.

In the first step of the cubane synthesis, Cole and Eaton got 85 percent of the possible theoretical yield. In subsequent reactions, they got yields of 30 percent to 98 percent. You might think that the main reason they wrote down these yields is to demonstrate efficiency. Indeed, it's easy to calculate how many carloads of the starting material they would have had to use to get one milligram of cubane, if each step were 10 percent efficient. But that's not the main reason these workers listed the percent yield.

The yield in a chemical reaction is an aesthetic criterion. To appreciate this, let's think about how one might get only a 10 percent yield. A reaction is a sequence of physical manipulations performed by a fallible human being using imperfect tools. One way to get a 10 percent yield is to spill 90 percent of the solution in the course of a transfer from a flask to a filter funnel. Sloppiness will not impress people, in science or art.

Suppose every transfer is meticulously done. The craftsmanship is high. Still one gets a 10 percent yield. Now it's not human hands that are at fault, it's the mind. Nature has paid no attention to our design but has decided to do something else with 90 percent of our material. That doesn't show control of mind over matter, it doesn't draw admiration. Maybe there's a better way to carry out that reaction step. A sequence of high-yield reactions, such as the cubane synthesis contains, defines elegance in chemistry.

There is high logic in synthetic strategy. The design of a multistep synthesis resembles the making of a chess problem. At the end is cubane—the mating situation. In between are moves, with rules for making them. The rules are much more interesting and free than those of chess. The synthetic chemist's problem is to design a situation on the chessboard, ten moves back, which has the most ordinary appearance. But from that position, one player (or a team of chemists), by a clever sequence of moves, reaches the mating position no matter what the recalcitrant opponent, the most formidable opponent of all, Nature, does.

John Cornforth, a great synthetic chemist himself, has made the wise point that this opponent (he calls it truth) "sometimes changes during the work into a teacher and friend."[2]

The obvious logical content of synthesis has inspired people to write computer programs to emulate the mind of a synthetic chemist. The design of such programs is a high challenge to researchers in "artificial intelligence" and "expert systems" as well as to chemists. The programming is an educational act of great value; chemists who have worked on these programs have learned much about their own science as they analyzed their own thought processes. Use of these programs is now common in some industrial laboratories—they can be of help in routine syntheses.

Can the synthesis programs suggest *interesting* syntheses, the kind that if turned to practice could be published in a good chemical journal? I think this remains an open question. The papers of workers in computer-assisted synthesis typically illustrate the capability of their programs by demonstrating that the programs suggest routes to difficult targets identical to ways conceived earlier by other good (noncomputerized) chemists. But I'm still waiting for the paper that begins:

> There is great interest in a new antiviral agent, Bussacomycin-F17, isolated from the slime mold *Castela manuelensis*. We attempted a total synthesis of this molecule with 15 asymmetric centers, but were unsuccessful. We then turned to the program, MAGNASYN-3, which suggested the eventually successful synthesis shown in Fig. 1 . . ."

The psychology of human beings is not well suited to admitting that we can be replaced by a computer program, only that others can be.

A chemical synthesis is obviously a building process. One therefore sees architectonic considerations and the aesthetics of architecture figuring prominently. Note, for instance, that intermediates in the synthesis of cubane are more complicated than the starting material or product. Why is this so? Well, scaffolding has to be built to hold pieces of the structure in place while other parts are assembled. A specific detail gives some further insight. In I (see illustration 21.1) there are two CO, or "ketone" groups. The reaction to II transforms one of these (the "top" one) to a five-membered ring, but leaves the other one alone. Then Cole and Eaton get to work on that other one, change it from CO to COOH (III→IV), the COOH to $(CH_3)_3COOCO$ (IV→V), that to H(V→VI). In VI→VII, they uncover the second ketone group and then proceed to do to it the same violence they did to the first one (VII→VIII→IX→X). What a waste of effort! Why not do both at once?

What you see here is the basic and simple idea of a "protecting group," the padding or concealment of one piece of a molecule while a transformation is done on another piece. Then the protecting group is removed. When cubane was first being made, Eaton and Cole were fearful that this molecular skeleton might be unstable. So they proceeded in cautious little steps, using this strategy of protection.

They need not have worried. As Phil Eaton told me, we know today that actually both CO or ketone groups can be converted to COOH in one step. That this was not attempted the first time the molecule was made in no way detracts from the synthetic achievement. It points out the "historicity" of this human activity, as all others: something was done, perhaps not as well as it could be done today, in tentative steps, but still created, for the first time, by human intelligence, human hands.

Synthesis is a building process, but what a marvelous "hands-off" kind of building! This is not the nailing together of a wood box shaped like a cube, or even a Palladian villa. In the reaction flask there is not one molecule but 10^{23}. They are tiny. They are all bouncing around, chaotically doing their own thing. And yet on the average they are being made to do what *we* want them to do, driven only by the external macroscopic conditions we impose on the flask and the strong dictates of thermodynamics. We create local order, to order, through an increase of disorder in the surroundings.

Here is what R. B. Woodward, a great synthetic organic chemist, wrote:

> The synthesis of substances occurring in Nature, perhaps in greater measure than activities in any other area of organic chemistry, provides a measure of the condition and power of the science. For synthetic objectives are seldom if ever taken by chance, nor will the most painstaking, or inspired, purely observational activities suffice. Synthesis must always be carried out by plan, and the synthetic frontier can be defined only in terms of the degree to which realistic planning is possible, utilizing all of the intellectual and physical tools available. It can scarcely be gainsaid that the successful outcome of a synthesis of more than thirty stages provides a test of unparalleled rigor of the predictive capacity of the science, and of the degree of its understanding of its portion of the environment.[3]

And E. J. Corey, a modern master:

> The synthetic chemist is more than a logician and strategist; he is an explorer strongly influenced to speculate, to imagine, and even to cre-

ate. These added elements provide the touch of artistry which can hardly be included in a cataloguing of the basic principles of Synthesis, but they are very real and extremely important. . . .

The proposition can be advanced that many of the most distinguished synthetic studies have entailed a balance between two different research philosophies, one embodying the ideal of a deductive analysis based on known methodology and current theory, and the other emphasizing innovation and even speculation. The appeal of a problem in synthesis and its attractiveness can be expected to reach a level out of all proportion to practical considerations whenever it presents a clear challenge to the creativity, originality and imagination of the expert in synthesis.[4]

Interestingly, it was Woodward who, through the verve and style of his syntheses, made chemists feel that "the art of synthesis" was high art indeed. And Corey has written a book entitled *The Logic of Chemical Synthesis*.

It might seem that in the making of things art and logic pull in opposite directions, forming still another axis. But in the service of creation, there is another kind of synthesis, of the twain.[5]

22. The Aganippe Fountain

At Millesgården, on the island of Lidingö near Stockholm, the work of the great Swedish sculptor Carl Milles is splendidly displayed. During a recent visit there I saw one sculpture group, the Aganippe fountain, in a new light. Its theme is classical in origin, but Milles's interpretation is idiosyncratic. The spring of Aganippe, on the slopes of Mount Helicon in Greece, was said to inspire artists and poets. Milles portrays Aganippe as a female figure, recumbent but in motion at the edge of the pool, and reflected in it. From the pool rise several dolphins, arched in mid-leap. Three of the dolphins carry on their backs men symbolizing Music, Painting, and Sculpture. Water rises from the beaks of the dolphins; this is, after all, a fountain, and Milles was a master designer of fountains (illustration 22.1).

The Aganippe sculpture group always gave me pleasure when it graced a courtyard in the Metropolitan Museum of Art in New York. It has now been moved to Brookgreen Gardens outside of Charleston, South Carolina. At Millesgården one sees a replica, containing somewhat fewer figures. It remains lovely.

Fountains are about water—its motions, divisibility, and reunion in flow. They are also about artifice—the real and the imagined, the natural and the unnatural. It is this last distinction that I want to explore, first by showing how the artist and scientist may confound this distinc-

Chapters 22–26 are adapted from Roald Hoffmann, "Natural/Unnatural," *New England Review and Bread Loaf Quarterly* 12, no. 4 (1990): 323–35. I am grateful to Emily Grosholz for her careful editing of this article.

22.1 The Aganippe
Fountain by Carl Milles at
Millesgården on the island
of Lidingö, near Stockholm.
(Photographs by the author)

tion for good reasons, and then by arguing that the distinction has
some warrant after all.

One of the mounted figures that rises from the fountain—a man,
balanced on a dolphin's back—represents Sculpture. He is life-size,
much larger than the stylized, diminished dolphin, and yet this dispro-
portion doesn't matter. The man is dancing, and gravity's pull is light
on him. Milles's art, his recurrent aim, was to defeat gravity. In bronze
sculpture! The water, which emerges as several thin jets from the dol-

phin's snout, is angled upward; it falls back, under the natural force of gravity, and sprays the young man. He reaches backward, and on one outstretched hand rests (that's not the proper word for Milles's sculpture; more precisely "is balanced") a horse. The horse is small, the size of the man's forearm, but it is real, and galloping in the air. On the horse's head, in final defiance of gravity, another, smaller man is balanced—flying, falling, flying (illustration 22.2).

What is natural and what is unnatural about this work, which is both a fountain and a sculpture? Like all fountains, it is patently synthetic, artificial, and unnatural. Someone has thought up a clever device, combining art and hydraulic engineering, to manipulate for aesthetic purposes one of the essentials of life and the earth, water. Fountains are sculpture with the unique feature that water is used as a sculptural element. And a substantial part of their interest derives from the fact that they overcome the tension of opposites between solid bronze or stone and moving, seemingly free water. How could these elements possibly be integrated? And yet in this kinetic sculpture they are.

22.2 Two details
of the Aganippe
Fountain
by Carl Milles.
(Photographs by
the author)

The artifice is that the water doesn't "want" to run up, nor does it want to run in controlled channels, much less through dolphins' beaks! We conspire to manufacture elaborate mechanisms to channel water, to pump it up so that it can flow down naturally and, in seeking its own level, in some places even to make it run straight up. Pumps, meters, gates, valves—God, all those hidden mechanics of the artificial! What could be more synthetic than a fountain?

The fountain's figures are cast in bronze, their mechanical elements made from other metals. The bronze itself is artificial. Or is it? Bronze is an alloy of copper and tin (perhaps with a little lead and zinc), an alloy of sufficient importance in the history of mankind to have an Age named after it. The alloy is both harder and more fusible than its component elements, which in turn are smelted from their ores, refined in a remarkable metallurgical process by men and machines. The ores of copper and tin—covellite, cuprite, cassiterite, and others—are certainly natural. But they have not always lain in the earth unchanged. They came into existence under the action of forces operating perhaps more weakly over a longer time (geochemistry) than human metallurgical intervention, or more strongly in a shorter time (the nuclear transformations in the early seconds of the universe).

Thus here in Milles's fountain, natural ores, unnatural smelting and alloying technology, are used by natural man in the patently unnatural act of sculpture to manipulate the most natural of elements, water, and to construct images of natural man and horse and dolphin. And these are all perceived by my biological eye as a fountain that pleases me and that I can compare with Roman fountains I've never seen except through their unnatural images on natural but manufactured paper! Any imagined separation of the natural and unnatural can be confounded in the examination not only of Milles's fountain but also in the careful analysis, aesthetic or scientific, of any object in our experience.

23. Natural/Unnatural

Scientists, especially chemists, will probably like the argument that ends the previous chapter. They often feel beleaguered by society because they produce "unnatural," and sometimes downright dangerous, materials. A cursory survey of the media shows a consistent use of negative descriptive terms whenever chemistry is mentioned. Adjectives such as "explosive," "poisonous," "toxic," and "polluting" are so closely conjoined with chemical names or nouns that they have become stock phrases. Whereas "natural," "organically grown," "unadulterated," and so on are given positive connotations, synthetics may seem at best conditionally good. Yet synthetic substances are widely manufactured and bought. For they do shelter us, heal us, make life easier, more interesting, and more colorful. Chemists encounter frustratingly conflicting signals from society—economic dependence and reward, coupled with an attitude from the media and some intellectuals that is abusive. I wonder if there are some parallels to the attitudes toward Jewish moneylenders in Europe in the Middle Ages.

One might advise chemists engaged in pure research not to take upon themselves the burden of guilt incurred by the often greedy, sometimes unethical producers and sellers of a dangerous chemical.

But that is a subject that deserves its own extensive discussion; rightly or wrongly (I think both), many chemists feel that the media and society are negative not only about businessmen but about chemistry and chemists.

One should also make the' distinction between the words *man-made* (or *woman-made*), *synthetic*, and *unnatural*. Common words are not insulated from the alternative meanings that usage has built for them. As one moves from *man-made* to *unnatural*, the number of such other meanings, with their associated negative connotations, clearly compounds. Nevertheless, I will use these words interchangeably, because I think they are so used in the dialogue around chemicals and people.

So the scientists will welcome what seems to me undeniable, that in any human activity—art, science, business, or childrearing—it makes little sense to separate the natural and unnatural. Both are inextricably intertwined, for there is an inherent ambiguity in any attempted separation.

An artist reflecting on his vocation will not object, in my experience, to valuation of the unnatural as a common creative link of art and science. Some artists have gone even further, as Igor Stravinsky does in his *Poetics of Music*. He inveighs against the idea that natural sounds are music, or that music should imitate nature:

> I take cognizance of the existence of elemental natural sounds, the raw materials of music, which, pleasing in themselves, may caress the ear and give us a pleasure that may be quite complete. But, over and beyond this passive enjoyment we shall discover music, music that will make us participate actively in the working of a mind that orders, gives life, and creates. For at the root of all creation one discovers an appetite that is not an appetite for the fruits of the earth.[1]

The chemist will go on, as I will, to make the point that all substances—water, bronze, the patina on that bronze, Milles's hands, my eyes—all have a microscopic structure. They are composed of molecules. The component atoms, their arrangement in space, confer upon these macroscopic substances their various physical, chemical, and biological properties. As we have noted, as subtle a difference as one molecule being the mirror image of another will affect whether it is sweet, or addictive, or a toxin. Much of the beauty of modern biochemistry is in the unraveling of the direct mechanisms of action of natural, biological processes—how precisely O_2 bonds to the hemoglobin in our red blood cells, and why CO binds better. That nylon replaced silk in women's stockings is not just a lucky circumstance—there are important similarities, on the molecular level, in the composition and structure of the two polymers (amide, carbonyl groups; pleated sheet structures; hydrogen bonding . . .). The singular intellectual achievement of chemistry in our time is the comprehension of the structure of molecules, covering the range from pure water to the alloy bronze or the protein rhodopsin in the cones of my eye.[2]

But lest the scientist feel too comfortable, I will go on to defend the distinction between "natural" and "unnatural." This division has good reasons for its historical persistence. No amount of supposed "rationality" will make the real intellectual concerns go away, and they persist for scientists as much as they do for other people.

In chemistry the natural/unnatural dichotomy has an interesting history. Early distinctions between organic and inorganic substances were swept aside by the demonstration, first by Hermann Kolbe in 1845 for acetic acid, that naturally occurring substances could be synthesized from entirely inorganic, inanimate sources.[3] Note the subtle difference of emphasis here—organic versus inorganic, not natural versus unnatural. Both organic and inorganic molecules required human manipulation to be shown to be identical.

The identity of substances remains to this day a subject of dispute and economic value. For instance, chemists will typically scorn health food stores advertising (and selling at a premium) rose-hip vitamin C, as being different from synthetically produced vitamin C. The same chemist, call him A, will get very upset when a colleague, B, says that a synthesis of some molecule reported by A cannot be reproduced. What has probably happened is that a reagent in one synthesis contains some adventitious admixture of a catalyst, due to its mode of preparation. That "dirty" catalyst made the reaction go for A but was absent from B's reaction flask. Pure vitamin C, synthetic, is identical to natural vitamin C. But a bottle of vitamin C made from rose hips is certainly not identical to a bottle of vitamin C made by a chemical manufacturer— at a parts-per-thousand level. I'm not implying that there are important differences, simply that there could in principle be differences in substances that are perforce impure and hence mixtures.

Chemists might reflect on the fact that despite the seeming irrelevance of the organic/inorganic and natural/unnatural divisions in chemistry, that in their own language and social structure the dichot-

23.1 Vitamin pills (Photo by Ken Whitmore/Tony Stone Images)

omy has a life of its own. For instance, people in the molecular trade talk of "natural product synthesis" (i.e., the synthesis of molecules found in nature) to distinguish it from the synthesis of molecules never present on earth before. But, significantly, no chemist uses the term *unnatural products* except as a joke. The slightly guarded humor of the phrase betrays, as humor often does, some of the ambiguous feelings chemists often harbor on this subject.

The chemist also distinguishes the discipline of biochemistry, which concerns the nature and mechanism of basic chemical processes in living organisms. Biochemists often aim to understand the mechanism of such processes by reducing them to a sequence of individual chemical actions. But the organic, inorganic, and physical chemists who study these individual fundamental steps would rarely be able to secure a job in a biochemistry department! Synthetic chemists speak with approbation of *biomimetic* methods (i.e., synthetic procedures imitating natural ones). The prefix "bio" obviously carries some psychological and social value. Such professional divisions and specializations persist in giving the natural/unnatural dichotomy life even within chemistry.

24. Out to Lunch

The personal conduct of scientists is also revealing. The scenario that follows is a pastiche of several recent experiences. Not long ago I was the guest at lunch of an executive of a major chemical company. We were at a luxurious, recently opened restaurant, proud of bringing la Nouvelle Cuisine to this corner of America. The chairs were of light wood, delicately caned, and the napkins felt like fine linen. One could see the touch of someone with training in Ikebana in the fresh flower arrangements.

I was prepared for small talk, platitudes, and some good science. Instead my host proceeded to unburden himself in an emotional tirade against a few young people, the American equivalent of the "The Green Ones" in Europe, who had given him a hard time at a press conference that morning.

These young people (he kept coming back to them) dominated the public discussion after he had presented a plan for building a new agricultural chemical manufacturing facility for pesticides and herbicides. They asked him if the chemicals to be produced were adequately tested for mutagenicity, and questioned the company's control of effluents. They reminded him, aggressively and, he thought, arrogantly, of a previous mishap at another facility of that company. He found their criticism full of fears, unscientific, and irrational. It seemed they doubted the need for the pesticide, a deterrent to weevils; they thought natural methods of pest control were adequate. The older man, a distinguished chemist and obviously a good businessman, was upset, perhaps because he could not allow himself to be upset at the press conference. He fumed about the confused anarchy of these people and also hinted at organized political motives. The good wines, first a New York State chardonnay and then a superb Saint Emilion, did settle him a little. After the white wine he was able to joke about the then current Austrian wine adulteration scandal (diethylene glycol, a component of antifreeze, had been used to "sweeten" wine). In time, the pleasure of telling a receptive visitor about a find he had made in an antique store, an unusual Indian basket (we shared an interest in Native American art), took over. After lunch we strolled in the gardens around the restaurant, admiring especially the purple and black tulips then in bloom.

25. Why We Prefer the Natural

One does not need a business executive and a fancy modern restaurant for this scenario. I suspect that the strongest defenders of the lack of a separation of the natural and unnatural go home to houses with picture windows and not with large photographic enlargements of exotic landscapes in place of those windows. In their homes grow real plants, not artful plastic and fabric imitations. No solarium will substitute for their real Algarve or Bahamas tan; they will avoid like the plague plastic shingles on their house and wood grain imitations in their dining room furniture; they will complain about what the European Economic Community is trying to do to their beer. It seems clear to me that the scientist or technologist who complains about other "unreasonable" people not being able to see the impossibility of separating the natural and the synthetic nonetheless testifies to the hold such a separation has on his or her own psyche in daily life.

So let us think about why it is that we prefer the natural, no matter who we are and what we do. I see many interconnected psychological and emotional forces at work—among them six that I can label: romance, status, alienation, pretense, scale, spirit.

Romance: In the second act of Tchaikovsky's opera, *The Queen of Spades,* a masque or pastorale, "The Faithful Shepherdess," is interpolated, which does not exist in Pushkin's original story. Daphnis and Chloe sing of the pleasure they take in nature, in a marvelous Mozartean duet (illustration 25.1).[1]

25.1 A piano and vocal score of a duet from the pastorale in act 2 of Tchaikovsky's *Queen of Spades*.

The tradition of the pastorale is as old as that of fountains. This kind of romance derives from an unrealistic striving for what no longer is or cannot be. The pastorale may even be a way to distance oneself from the pastoral—the irony of these unreal and unnatural but entrancing constructions supposedly about the natural is that pastorales were fine for everyone except the people who had to make a living in the pasture. The royal courts are gone, but romantic traditions persist. A reaching out for nature, for real wood, the smell of hay, the feel of the wind in the sails, still determines our *desires*. It doesn't matter that the real stable smelled bad, or that train stations were dirty, noisy structures. I see Ingrid Bergman saying goodbye to Leslie Howard at the train station, and I know all train stations. I feel them within me. My mind's stable smells just right.

Status: The real success of the synthetic is due to either some combination of lower cost, greater durability, more versatility, or even new capabilities, relative to some natural materials. This is the polymer century, when large synthetic molecules have replaced one natural material after another: nylon in place of cotton in fishing nets, fiberglass instead of wood in boat hulls. The replacement or new use (polyethylene as a food wrap, for example) is invariably a democratizing process, for a wider range of materials is made available more cheaply to a larger group of people. Sanitary water delivery and waste disposal, a wider spectrum of color, better shelter, fewer deaths in childbirth and infancy are now available to many more than those who could enjoy such luxuries and essentials a hundred years ago.

But human beings are (nicely) strange. When they have some of anything, they want more. Or they simply want something better than their neighbors. When the synthetic becomes inexpensive and available to all, a curious inversion of taste occurs: the arbiters of elegance decree that the "natural" carries more cachet. If a cotton shirt is supposed to feel more luxurious than a "permanent press" blend, sure enough the shirt begins to feel that way. A wood floor is certainly perceived as being nicer than linoleum, and the rarer the wood the better.

Perhaps I've been too negative here. Perhaps silk does "feel" nicer than nylon. Perhaps what we want is not so much to be superior to someone else as to be somewhat (not too much!) different. The natural provides, in its infinite variability, that opportunity to be slightly different.

Alienation: We are becoming distanced from some of our tools, and from the effects of our actions. We see it in routine work on an assembly line, in selling lingerie, even in scientific research. We work on a piece of something, not the whole. To be efficient we work repetitiously, so that we may even lose interest in the whole. Mountains of paper insulate us from the human beings affected by our actions. Around us proliferate machines whose workings we don't understand. I doubt that there are many among my colleagues who could do what Mark Twain's Connecticut Yankee in King Arthur's Court could do— that is, to reconstruct our technology from all those partial differential equations we know. We press buttons and elevators come (or don't come). Worse, we press buttons and missiles are launched, and only the victims see the blood.

The synthetic, artificial, and unnatural is almost always a factory-produced multiple, inexpensive because mass-produced. To be mass-produced it must be stamped, cast, or pressed repeatedly. The objects so made appear identical. In principle their design could be good, in practice design is sacrificed for economy. The typical mass-produced object shows little of the history of its making, neither in design nor in execution. Tetracycline antibiotics, for instance, are isolated from a culture of living organisms, chemically modified, purified, and packaged by remarkable, inventive tools and devices. But a bottle of fifty tetracycline pills hides the ingenuity behind that multiple product, its manufacture by human beings using tools of their own design.

There is something deep within us that makes us want to see the signature of a human hand on a product. There are clever ways to individualize mass-produced items. I think of the color variations on the prints of F. Hundertwasser (hardly inexpensive) or the cheerful ceramics that Stig Lindberg designed for Gustavsberg in Sweden in the 1950s. They should be encouraged.

Pretense: The false has a negative connotation in all things of significance to human beings. To tell a lie, to pretend to be what one isn't, is not to be good. Much of the synthetic world of chemicals is not only unnatural in the sense of being man-made, it also often pretends to be what it is not. In part this is a natural consequence of replacing something familiar with something else not very different, but stronger, more resistant to heat, and so on. So plastic plates carry the patterns of porcelain, and sheet plastic in furniture often imitates the wood grain. Napkins emulate linen, lace, and embroidery. There is the ancient profession of marbling. I was once told by a young man who was apprenticed in this honorable craft that to be good at it one should not only study marble but also think, while painting, of the geological forces that shaped it. Now some of this is fine, but too much imitation, a dissembling that accumulates, inevitably leads to revulsion. One longs for the authentic, the real.

25.2 A classical porcelain pattern on a plastic plate, a paper doily imitating lace, a paper napkin patterned after a Persian tapestry.

Scale: There can be too many of one thing, and too many, period. The first plastic ashtray, or titanium jewelry, looks interesting, but as more and more of them invade our environment, they quickly begin to bore us. The repetitive nature of its production (the key to its economic success) is often the only feature that a mass-produced object stylistically communicates to us.

Sometimes the very superabundance of artificial objects in our environment, rather than the repetition of one and the same one, repels us. The typical American motel room, for instance, offers us little respite from the artificial. The variety of plastics and synthetic fibers in the furnishing of such a room is astonishing and even intellectually interesting, as an exemplar for a course about polymers, or in thinking of the problems such a room will pose for future archaeologists. But one is hardly attracted to that setting.

Spirit: What makes scientists, indeed all of us (because scientists are no different from other people), seek out the natural? No simple psychological or sociological explanation suffices.

A perceptive scientist, Jean-Paul Malrieu, has written:

The linen cloth is something we share, at least in an imagined way, with our grandparents and remote ancestors, with heroes, with history. And this is a noble and valuable feeling. We belong to a long stream, and we remember, we do not rush to the final sea. The same for wood, stone— their everyday contact reminds us of other forms of life, of times in the history of the earth when mankind was not yet announced; the pottery on our shelves speaks to us of elsewhere, of other tribes, other needs. And of clay.[2]

A genetic and evolutionary argument for the strong affinity of human beings for the living world has been provided by Edward O. Wilson in his biophilia hypothesis.[3] This rings true to me.

Laura Wood, an early reader of my manuscript, points out that people have such strong feelings on environmental issues because "for some this is a deeply spiritual issue. . . . Since matter is imbued with spirit, the world itself is sacred and to be treated with respect."[4]

I believe that our soul has an innate need for the chanced, the unique, the growing that is life. I see a fir tree trying to grow in an apparent absence of topsoil, in a cleft of a cliffside of Swedish granite near Millesgården, and I think how it, or its offspring, will eventually split that rock. The plants trying to live in my office remind me of that tree. Even the grain in the wood of my desk, though it tells me of death, tells me of that tree. I see a baby satisfied after breast feeding, and its smile unlocks a neural path to memory of the smiles of my children when they were small, to a line of ducklings forming after their mother, to that tree. As A. R. Ammons says, "My nature singing in me is your nature singing."[5]

26. Janus and Nonlinearity

What about the sinister side of that Janus-like image of chemistry? Actually, I think the public's view of chemistry is not as bad as chemists think, if you come to terms with the fact that human beings are nicely nonlinear in their thinking—they can hate and love, stand in horror of and value one and the same thing. I remember the way chickens were killed in back of our house in Poland, and I still shiver at the memory. I love chicken dishes, but I don't want to see the chickens killed. Or take attitudes toward doctors—I grew up in an immigrant middle-class Jewish family where all the parents wanted their children to be doctors. Yet if you would listen to what they said about doctors, you'd hear an uninterrupted litany of complaints—doctors misdiagnosed, they were inhuman, they were only interested in money, and so forth.

Many people are afraid of chemistry; yet the same people, not any different ones, also value chemotherapy and polyethylene. So (and here I speak to my fellow chemists) when you are attacked by seemingly irrational environmentalists, I want you to take a deep breath, slow the angering rush of blood, open up your hearts. No one is attacking you. The environmentalist, the one who doesn't want our nest fouled, is you too. I hate to see human beings polarized, by religion, race, or politics. It is not "us" (whoever "us" is) versus "them," those irrational, Luddite critics of our lifestyle. There is so much of "them" in "us"—allow for that life-enhancing and beautiful complexity of human beings, a complexity that does not forbid a chemist to be incensed at a rotting chemical dump at the same time as he or she knows that the production of those chemicals increased our life span.

PART FOUR

When Something Is Wrong

27. THALIDOMIDE

Chemie Grünenthal was one of many small pharmaceutical firms in postwar Germany. It first made antibiotics for other companies, but in the 1950s the firm ventured into its own modified penicillins. The German drug market was quite open then; neither the efficacy nor safety of a drug had to be proven in great detail. Almost anything was available over the counter, and the success of a product depended as much on advertising and marketing as on its value.[1]

It was in the 1950s that Valium and Librium were introduced, tranquilizers which were an instant success. Illustration 27.1 shows the structure of diazepam (Valium) as well as barbital (Veronal), a com-

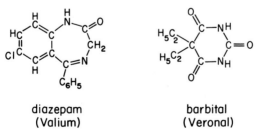

diazepam
(Valium)

barbital
(Veronal)

27.1 The structure of diazepam (Valium; *left*) and barbital (Veronal; *right*).

Much of the information in this chapter is drawn from Henning Sjöström and Robert Nilsson, *Thalidomide and the Power of the Drug Companies* (Harmondsworth: Penguin, 1972).

mon barbiturate sedative. It was natural for pharmaceutical companies to explore compounds that were chemically similar, even in the vaguest way, to these molecules. There was a lot of money to be made in the sedative and tranquilizer market.

Given its size, Chemie Grünenthal had only a small scientific department, headed by a physician, Dr. Heinrich Mückter. In 1954, Wilhelm Kunz, a chemist on his staff, by training actually a pharmacist, synthesized (N-phthalidomido)-glutarimide ("thalidomide"), the molecule whose structure is shown in illustration 27.2. Note the superficial resemblance to the sedatives shown above. Note also the presence in thalidomide of a carbon with four different groups around it (marked by an asterisk in illustration 27.2), pointing to the existence of enantiomers, nonsuperimposable mirror images. The molecule as used medically was a mixture of the two enantiomers.

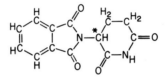

thalidomide

27.2 The chemical structure of thalidomide.

Driven by the resemblance I've pointed to, the Chemie Grünenthal researchers convinced themselves that the molecule had good sedative properties. The reason I say it in this way is that subsequent investigations have failed to confirm the sedative qualities claimed. The toxicity of thalidomide was low, and this encouraged the manufacturers to put the drug on the market. It was first introduced as part of a drug combination directed toward respiratory infection in 1956 and shortly thereafter sold as a sedative directly and in dozens of combinations in Germany.

The company needed published articles testifying to the utility of the drug. So it sought them out. In the Grünenthal files there is a report from their Spanish representative that a certain doctor "had declared he was prepared to write a short report on Noctosediv [the Spanish trade name for thalidomide] whereby he would leave it to us to revise the final draft." In the United States, in 1959, Ray O. Nulsen, a Cincinnati physician, was convinced by Dr. Raymond Pogge, the medical director of Richardson-Merrell, the American company that was try-

ing to market thalidomide under license from Grünenthal, to "test"
the drug. Here is part of Nulsen's deposition in a subsequent trial
(Spangenberg is an attorney taking the deposition before the Eastern
District Court of Pennsylvania):

> "I note, doctor," Spangenberg said "that he (Dr. Pogge) asked you to
> start testing promptly and to send in reports. Do you have copies of the
> reports you sent in?"
>
> "No, it was all verbal," Dr. Nulsen replied.
>
> Dr. Nulsen later said he had passed on the testing information to Dr.
> Pogge "by telephone, or it may have been that we had lunch together, or
> it may have been when we played golf." . . .
>
> This information was eventually collected in an article published un-
> der Dr. Nulsen's name in the June 1961 issue of the *American Journal
> of Obstetrics and Gynecology,* entitled "Trial of Thalidomide in Insomnia
> Associated with the Third Trimester." This rather detailed publication
> put forward the conclusion: "Thalidomide is a safe and effective sleep-
> inducing agent which seems to fulfil the requirements outlined in this
> paper for a satisfactory drug to be used late in pregnancy."
>
> Spangenberg: "Who wrote the article, Dr. Nulsen?"
>
> Dr. Nulsen replied, "Dr. Pogge. I supplied him with all the infor-
> mation."
>
> At another point the attorney asked, . . . "your article cites about half
> a dozen German magazines and German texts. [Dr. Nulsen did not read
> German] Did you ever read these articles?"
>
> Nulsen: "No. That was supplied to me."
>
> Spangenberg: "You also cite Mandarino, another doctor, and foot-
> note the citation, and the footnote reads, 'To be published.' Did you
> ever see his article?"
>
> Nulsen: "I don't remember having seen it."[2]

It turns out that in fact thalidomide is safe in the third trimester of
pregnancy. But the quality of the research cited here was at the time
unfortunately typical of the work of Chemie Grünenthal and its associated companies.

Henning Sjöström and Robert Nilsson, who have participated ac-
tively in the legal processes around thalidomide, cite another case in
their devastating book:

> During the early part of 1961 the Stolberg company [Chemie Grünen-
> thal] was told of a Dr. Davin Chou in Singapore who had successfully
> used thalidomide for the treatment of pregnant women. No details were
> given about the stage of pregnancy treated, the dosage used or the fre-
> quency of therapy. Finally, and most significantly, the brief report was

concerned only with the effect on the pregnant women themselves, and no mention was made of any possible effect on the fetuses. This lack of any specific detail did not deter Dr. Werner [a director of Grünenthal's medical-scientific department] from distributing a circular letter to "co-workers throughout the world" saying, "In a private clinic in Singapore Softenon [thalidomide] was given to pregnant women who tolerated the drug well."[3]

In 1958, Dr. Augustin P. Blasiu in Munich published an article in *Medizinische Klinik* in which he said, "Side effects were not observed with either mothers or babies." He had administered thalidomide to 370 patients, but only to nursing mothers. Chemie Grünenthal sent a letter to 40,245 physicians citing Blasiu's work, describing thalidomide as a drug "which does not damage either mother or child."[4]

In 1959 reports began to come in about severe neurological damage, neuritis, caused by thalidomide. These were steadfastly denied, obfuscated, and concealed by the Grünenthal people; and numerous attempts were made to stifle public reporting of these symptoms. Worse was to come.

In 1960 physicians in Germany and Australia noticed a striking incidence of a peculiar malformation in newborns. It was phocomelia, a deformity in which the hands are attached to the shoulders, and feet to hips, like the flippers of a seal (hence the name of the syndrome: Greek *phoke* = seal, *melos* = limb). The anomaly was sufficiently rare prior to that time (estimated incidence: one case in four million births) that most physicians had never seen a case.

These were not the only symptoms. To quote a Canadian study of the mothering of thalidomide children:

Limb-deficiency, though the most common and most striking anomaly, constituted only one element of the syndrome among a host of other deformities. The major external defects were coloboma (a defect in one or both eyes), microtia (smallness of the external ear) associated with partial facial palsy, depressed bridge of nose, and hemangioma (tumor) on forehead, cheek, or nose. The internal defects were found in the cardiovascular system, urogenital system, and intestinal tract; there were also abnormal lobulations of liver and lungs, dislocated hips, syndactyly (fusing of fingers or toes), horseshoe kidney, bicornuate uterus, atresia (closure of a normally open channel in the body), and absence of the gall bladder.[5]

27.3 Goya brush and wash drawing, *Mother Showing Her Deformed Child to Two Women*. In the collection of the Louvre, reproduced by permission.

Goya, that prescient explorer of the dark side of our world, drew a "natural" case of phocomelia. This is shown in illustration 27.3.[6]

Approximately eight thousand children were born alive with phocomelia or related abnormalities. Most of these were in Germany and England, but there were cases reported in some twenty countries. Only after the evidence mounted so that it could not be excused or argued away and after exposure in the press did Chemie Grünenthal withdraw the drug from the German market, in November 1961. The various drug companies around the world who licensed the drug followed suit, unconscionably slowly.[7]

Did thalidomide cause the terrible abnormalities observed? Animal testing done *after* the disaster clearly showed the teratogenic (malformation causing) nature of the drug. Thus monkey tests at Pfizer showed *every* embryo deformed when the mother was given thalidomide in a certain early period of pregnancy.[8]

Do you want another kind of proof? Examine illustration 27.4,[9] which shows the incidence of thalidomide-type birth malformations in Germany and thalidomide sales, both "normalized" to the same value at their highest point.

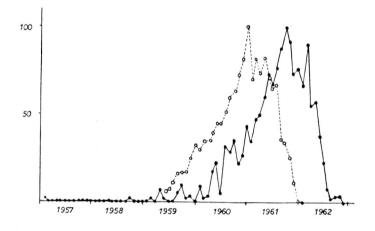

27.4 Incidence of thalidomide-type malformations (solid line, normalized to 100 in October 1961) and thalidomide sales (dashed, normalized to 100 in January 1961). Reproduced by permission of Churchill-Livingston Ltd.

Now we must face some of the obvious issues raised by this terrible story.

1.

Is thalidomide a *chemical* disaster? There seems to be only one chemist in the story, Wilhelm Kunz. Significantly, he was not one of the defendants in the ultimately futile legal process (1967–1970) instituted against Grünenthal in Germany, a trial vitiated by a compensation settlement between the company and the parents of the "thalidomide children." Of the seven principals in the case, five were physicians. The company's deceit of the public was mainly managed by doctors—*and* by the owners and management of the firms involved. So why blame this on chemistry?

I think there are two reasons for chemistry to share in the guilt. Thalidomide is a chemical. Chemists like to lambaste the public for its ignorance of the distinction between the natural and the unnatural. Indeed they are right to do so. But once you have taught people the chemical nature of all matter, and that the natural may sometimes hurt you, you must not try to hide that the synthetic also may, sometimes, hurt. This chemical did harm.

The worldwide public now has a variety of "chemical disasters" to choose from. There was the catastrophe of Bhopal in India (and there will be another one). There are tank cars of benzene or chlorine that are derailed. There is DDT; there are chlorofluorocarbons. There was mercury poisoning in Japan as there is currently in Brazil. I could have discussed any of these. In each case one could argue for an exculpation—this or that is not chemistry. Or even for a positive role for chemistry; who but chemists F. Sherwood Rowland and Mario J. Molina found the connection of chlorofluorocarbons to ozone depletion?

In each case economics and its dark side, greed, are dominant. But if chemists take credit for the positive trade balance contributions and the ulcer drug Tagamet, we also have to be willing to accept the blame. At least part of it. No chemist at Grünenthal (or another company) voiced any public doubts about the firm's behavior as reports of harmful effects just flooded in. No one blew a whistle. Only other physicians, and a free (and, yes, sensationalist) press, did.

There is another curious chemical angle to the thalidomide story. The molecule has one carbon in it with four different substituents around it. So thalidomide is chiral; this means, as we saw, that it exists in nonidentical handed forms. The reaction that produced it initially gave equal amounts of the left- and right-handed forms. And so it was used. There is some indication (contended) that the two enantiomers differ vastly in teratogenicity. The matter is somewhat complicated by the fact that the "harmless" enantiomer converts into the "harmful" one under physiological conditions.[10] The world is never simple . . .

Clear cases of differing biological activity of mirror-image forms of one and the same molecule abound. D-penicillamine is widely used in the treatment of Wilson's disease, cystinuria, and rheumatoid arthritis. Its optical isomer gives severe adverse effects.[11] The enantiomer of a tuberculosis drug, ethambutol, can cause blindness. Disastrous side effects associated with the painkiller benoxaprofen might have been avoided if the drug had been sold in its one-handed form. Such cases have led to regulatory pressure encouraging testing of pharmaceuticals as pure enantiomers. Some chemists and drug companies have resisted this, but others have realized there is creativity in designing handed syntheses. And profit.[12]

The twenty-five top-selling prescription drugs in the United States sold for $34.4 billion in 1993. Of these sales, 25 percent were of molecules that are not chiral, 11 percent were marketed as a mixture of the two mirror images, and 64 percent as one enantiomer. The handed category has been increasing at the expense of the mixture of enantiomers; in time all chiral drugs will be sold in a single, handed formulation.[13]

2.

Surely this (the behavior of Grünenthal, Richardson-Merrell Distillers, Astra, Dainippon, and others) is just bad science, isn't it? Just look at those quotes from the Spanish, American, and Chinese doctors! Were the science (pharmacology, biology, medicine, chemistry) done well, or at least just plain adequately, this would not have happened.

Indeed animal testing for teratogenicity of new drugs was routine in the major pharmaceutical companies. Hoffmann-LaRoche's Roche Laboratories published a major reproductive-system study of its Librium in 1959. Wallace Laboratories did so for Miltown in 1954. Both instances antedate the thalidomide story. Dr. Frances Kelsey, the FDA physician who courageously resisted enormous Richardson-Merrell pressure to license thalidomide in the United States, had good reasons for her reluctance. As a student in 1943 she demonstrated (with F. K. Oldham) that the rabbit fetus could not break down quinine, while the liver of the adult rabbit does so effectively.

The answer is twofold, I think. First, yes, this is abysmal science. And whereas science as a system for gaining reliable knowledge works in spite of instances of poor-quality experimentation—it will easily survive sloppiness, hype, and even fraud—the kind of science that touches on human lives cannot afford to be bad. The thalidomide disaster should not have been allowed to happen. Yet not a single drug company (all that competition in the sedative market!), not a single individual screamed, either before, or during, while thousands of adults suffered from neuritis and children were born into a less than fully human life.

The system failed. Science and medicine (chemistry in part) failed. The remedy had to come from legislation on drug testing that slowly was introduced around the world in the 1960s.

The second answer, I think, is, yes, this is bad science. But it isn't just bad science. It is an insidious failure of the system, because it intersects with the banality of evil, in Hannah Arendt's famous phrase. None of these people—the unethical doctors who supplied the results the company wanted, the salesmen in the field, the manipulators and distorters of data such as Dr. Mückter, the lawyers, men who threatened legal action against the physicians who first reported side effects, the medical journal editors who stalled on publication because a company objected—none of these were just plain evil. I'm sure they were good but flawed men (maybe even a woman here and there) who each in their own little way heard and saw something, but then in the gray area between doubting (as they should have) or following company policy, in that moral neither here nor there, neither black nor white land, they chose just a little here, just a little white. They then passed along a selected, slightly distorted message to human beings of similar moral weakness, who massaged the data just a little bit more, ignored what they didn't want to see, refused to read that memo with the bad news in the file, attributed that reaction to hysteria.

3.

So the outcome was bad. Then legislation cured the problem—but now haven't we gone overboard? The creativity of the drug designer is stifled; it costs $100 million to bring a drug to the market just because of all those mandated safety and efficacy tests. The net results, so this argument goes, is that we have kept more drugs from the market and thereby indirectly lost the lives of many people.

When I hear this argument, I am tempted to do what I do not want to do, and show the picture of a thalidomide child. It is not a matter of how many lives might have been lost because of the more stringent rules preventing new drugs, but of how many lives have been saved because the regulations prevented more thalidomide-like disasters.

If there be a calculus of risks and benefits, then the weighting that is applied to a single drug-induced phocomelia birth is (to me) so great that it outweighs any life or hundreds of lives saved. The anguish of the eight thousand thalidomide children and their parents is unimaginable. Nothing in the world can justify this. It must not happen again!

The theme of the same and not the same arises so clearly in the thalidomide story. The makers and sellers of the drug chose to see its resemblance to other sedatives and tranquilizers. They chose not to see the possibly different effectiveness and toxicity of the mirror-image forms of the molecule.

Primo Levi, in his wonderful autobiographical history, *The Periodic Table,* tells the story of an explosion he had while doing some research at the University of Turin. He needed sodium to dry an organic solvent, but he used instead potassium, another alkali metal, right under sodium in the periodic table. He writes of what the experience meant to him:

> I thought of another moral . . . and I believe that every militant chemist can confirm it: that one must distrust the almost-the-same (sodium is almost the same as potassium, but with sodium nothing would have happened), the practically identical, the approximate, the or-even, all surrogates, and all patchwork. The differences can be small, but they can lead to radically different consequences, like a railroad's switch points; the chemist's trade consists in good part in being aware of these differences, knowing them close up, and foreseeing their effects. And not only the chemist's trade.[14]

28. The Social Responsibility of Scientists

There are no bad molecules, only negligent, or evil, human beings. Thalidomide seems as harmful as they come, in the first trimester of pregnancy. But there have been persistent hints of its utility in treating inflammation associated with leprosy. And there are recent studies claiming that thalidomide can inhibit the replication of HIV-1 (the virus that causes AIDS).[1] Nitric oxide, NO, is an air pollutant but also an absolutely natural neurotransmitter. Ozone serves an essential (to us) function in the stratosphere, a thin layer of it absorbing much of the sun's harmful ultraviolet radiation. At sea level the very same molecule is a bad actor in photochemical smog, the atmospheric pollution caused mainly by automotive exhausts. Ozone destroys automobile tires (weak vengeance), plant life, and our tissues.

Molecules are molecules. Chemists and engineers make new ones, transform old ones. Still others in the economic chain sell them, and we all want them and use them. Each of us has a role in the use and misuse of chemicals. Here is what I see as our social responsibility as scientists to our fellow human beings.

I see scientists as actors in a classical tragedy. They (we) are sentenced by their nature to create. There is no way to avoid investigating what is in or around us. There is no way to close one's eyes to creation or discovery. If you don't find that molecule, someone else will. At the same time, I believe that scientists have absolute responsibility for thinking about the uses of their creation, even the abuses by others. And they must do everything possible to bring those dangers and abuses before the public. If not I, then who? At the risk of losing their livelihood, at the risk of humiliation, they must live with the consequences of their actions. It is this duty that makes them actors in a tragedy and not comic heroes on a pedestal. It is this responsibility to humanity that makes them human.

PART FIVE

How, Just Exactly, Does It Happen?

29. MECHANISM

The first, primeval activity of chemists is to respond to the question, What are you? After you know *what* you have, you next want to know *how* it came about. A natural question for curious human beings—whether coming upon the scene of an accident or seeing a gram of a strange, new molecule.

What is a mechanism? It is a sequence of irreducibly simple elementary chemical acts by which one molecule is transformed into another one. In a sense, it is like a computer program that describes how A goes to B, or a recipe that tells you how flour, eggs, sugar, butter, and chocolate chips are turned into cookies. It is also a history, of an action in the past but one that is repeatable today. The very word *mechanism* signals adherence to an analytical philosophy—assuming as it does a Newtonian clockwork operation of the universe, where whatever happens must be "explained" by a sequence of mechanical actions into which we, in the simplicity of our minds, divide up the continuous workings of nature.

Here is one study of a mechanism. In the early 1960s, Okabe and McNesby, at the then National Bureau of Standards (now called NIST, the National Institute of Standards and Technology), looked at the way

Professor Butts steps into an open elevator shaft and when he lands at the bottom he finds a simple orange squeezing machine. Milkman takes empty milk bottle (**A**), pulling string (**B**) which causes sword (**C**) to sever cord (**D**) and allow guillotine blade (**E**) to drop and cut rope (**F**) which releases battering ram (**G**). Ram bumps against open door (**H**), causing it to close. Grass sickle (**I**) cuts a slice off end of orange (**J**)–at the same time spike (**K**) stabs "prune hawk" (**L**) he opens his mouth to yell in agony, thereby releasing prune and allowing diver's boot (**M**) to drop and step on sleeping octopus (**N**). Octopus awakens in a rage and, seeing diver's face which is painted on orange, attacks it and crushes it with tentacles, thereby causing all the juice in the orange to run into glass (**O**).

Later on you can use the log to build a log cabin where you can raise your son to be President like Abraham Lincoln.

29.1 One kind of mechanism, by Rube Goldberg. Goldberg drew a series of inventions of Professor Lucifer Gorgonzola Butts, A.K. This one is an "Orange Juice Squeezing Machine." (Reprinted with special permission of King Features Syndicate.) Goldberg studied some chemistry in the College of Mining at the University of California, Berkeley.

ethane is broken down photochemically. This important hydrocarbon is on the left side in illustration 29.2. The symbol (hν) above the arrow is for a photon, for light. In this case not just any light but radiation very far into the ultraviolet region of the spectrum. Under the action of such energetic light, ethane gives ethylene plus a hydrogen molecule.[1]

The job of analysis is to determine that you have ethylene and not ethanol and not cholesterol. We do have ethylene and hydrogen, and nothing else. Now we want to know how that reaction really happens.

The study of mechanisms of chemical reactions is a textbook case for the application of the scientific method. You have an observation. You form several alternative hypotheses explaining that observation, and you proceed to eliminate the hypotheses (through experiment or

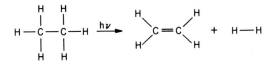

29.2 The photolysis of ethane yields ethylene and hydrogen molecules.

theory, but mainly through experiment), one by one, until you are left with one and that one must be right.

The first step in this protocol is to compile the list of possible hypotheses that would explain this observation. While I would not expect anyone who is not a chemist to write down many such possibilities, I would anticipate a professional chemist to write down two or three. There is one mechanistic hypothesis that I *would* expect everyone to write down without knowing anything about a chemical reaction at all. This is the "one-fell-swoop" or "by magic" hypothesis, attributing to nature the weakness of our own minds by assuming that it does everything all at once. This is sometimes dignified in organic chemistry by being called "a concerted reaction."

Illustration 29.3 shows this mechanism as the first of several. Since there are two carbons in ethane with hydrogens next to each other, why not have those two hydrogens come off together, to give in one step ethylene plus hydrogen? Well, that is one possibility. The second and third possibilities, esoteric to a nonchemist, derive from much previous chemical experience. The two hydrogens could come off from the same carbon to give H_2, the hydrogen molecule, leaving the C_2H_4 fragment with the right number of atoms, but not connected up in the right way. One of the carbons has three hydrogens, the other one has one hydrogen attached to it! So one would have to postulate, to complete this mechanism, a rapid step by which a hydrogen moves from one carbon to another (and there is precedent for such) to give the ethylene molecule.

The third mechanism is called a chain reaction. We know that light or heat or other forms of energy break bonds, usually one at a time, in molecules. So a chemist might postulate that the light would break a carbon-hydrogen bond to give a hydrogen atom and the remnant, which is called an "ethyl radical," C_2H_5.

Now you have to put yourself into the life of a molecule. These tiny

Hypothetical mechanisms:

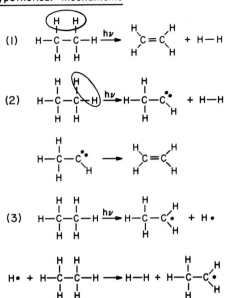

29.3 Three mechanisms for the photolysis of ethane.

little things are a fraction of a fraction of a fraction of a millimeter across. In any flask there are 10^{20} of them, hundreds of billions of billions, floating and bouncing around madly, constantly colliding with each other. The hydrogen that is knocked off by the light does not sit still. It is impelled by the energy that gives it freedom and by collisions with its neighbors to fly over to one of the 10^{20} other ethane molecules flying/floating around. A collision ensues. One of the things that we've learned hydrogen atoms do readily is that they abstract or pull off atoms from other molecules. So you could imagine the initially freed hydrogen during that collision pulling off another hydrogen atom from the neighboring molecule, to form the product H_2 molecule. We still do not have the ethylene product, but by a sequence of subsequent steps (which are not outlined here), one can actually also get the C_2H_4.

Next in order are experiments to eliminate, one by one, the mechanisms. The study of mechanisms in the last fifty years has been made much easier by the ready availability of isotopes. Recall from chapter 8 that isotopes are modifications of an element, which are different enough so that we can tell that they are there, but not different enough to matter (i.e., influence the reaction), as a first approximation. They are the ultimate spies . . .

The isotopic tracers of use in probing the mechanism of the ethane reaction were those of hydrogen, particularly deuterium, or "heavy" hydrogen. Okabe and McNesby took a mixture of normal ethane (C_2H_6) and an ethane in which every last hydrogen was replaced with deuterium (C_2D_6). Where did they get the "deuterated" compound? They bought it, and back in the laboratories of Merck it was synthesized. What was the first thing they did after they got a sealed flask or ampoule of the gas from the chemical supplier? They probably analyzed it. In this business you do not trust anyone. Someone just might have slipped up, might have put in five deuteriums instead of six.

You can see here how the study of mechanisms is intimately tied up with synthesis, and with analysis.

So the National Bureau of Standards researchers took this C_2H_6 and C_2D_6 mixture and photolyzed it. Let us trace the expectations of the various mechanisms. (Please refer back to illustration 29.3.) Mechanism (1) would give H_2 if the light were absorbed by C_2H_6, and D_2 if the light hit C_2D_6. Both are equally likely. No HD would be produced. Mechanism (2) would give H_2 from C_2H_6, D_2 from C_2D_6, and no HD at all. H_2, D_2, and HD are the three possible forms of the hydrogen molecule. They are of different weight: HD has a mass one and a half times that of H_2, while D_2 is two times as heavy as H_2. With a mass spectrometer, an inexpensive tool described briefly in chapter 3, they are easily told apart from each other.

Mechanism (3) is different. Supposing the light is absorbed by an ethane which has normal hydrogens, and knocks off a hydrogen atom. That hydrogen atom, once it is formed, wanders over (very quickly) to any of the billions of billions other ethane molecules. It could just as well hit C_2H_6 as C_2D_6 and it wouldn't know the difference. From the former it might pull off a hydrogen to give H_2, but from C_2D_6 the hydrogen atom would pull off a deuterium to give HD. Repeat the process in your mind beginning with a D atom breaking off from C_2D_6. Then imagine a pot full of equal amounts of C_2H_6 and C_2D_6. You can see that formation of HD would statistically happen more often than the formation of H_2 or D_2.

The experimental result Okabe and McNesby obtained (illustration 29.4) was unambiguous. In the photolysis of a mixture of C_2H_6 and C_2D_6 there is a preponderance of H_2 and D_2, and very little HD. Mechanism (3) is eliminated; it would have predicted much HD.

Experiments:

A) mixture of

$$H-\underset{\underset{H}{|}}{\overset{\overset{H}{|}}{C}}-\underset{\underset{H}{|}}{\overset{\overset{H}{|}}{C}}-H \quad \text{and} \quad D-\underset{\underset{D}{|}}{\overset{\overset{D}{|}}{C}}-\underset{\underset{D}{|}}{\overset{\overset{D}{|}}{C}}-D$$

yields mainly H_2 and D_2, only a little HD

B) photolysis of

$$H-\underset{\underset{H}{|}}{\overset{\overset{H}{|}}{C}}-\underset{\underset{D}{|}}{\overset{\overset{D}{|}}{C}}-D \quad \text{yields mainly } H_2 \text{ and } D_2, \text{ little HD}$$

29.4 Two experiments to elucidate the ethane photolysis mechanism.

Next experiment: The researchers then paid a bit more money to get a molecule which had not *all* the hydrogens replaced by deuterium, but exactly *half* of them (H_3CCD_3). For H_3CCD_3 photolyzed, mechanism (1) would give HD and only HD. Mechanism (2) would give D_2 or H_2, depending on whether the light "hit" the left or the right side of the molecule. Neither the light nor the molecule actually cares much about that left/right distinction. The experimental result is equally clear. Okabe and McNesby obtained primarily H_2 and D_2, and little HD. Mechanism (1) is eliminated. And therefore mechanism (2) is proven.

Or is it? Now we come to the workings of the scientific method and the role of human psychology. Of course mechanism (2) is not proven. You only falsify or disprove hypotheses, eliminate mechanisms—you do not prove them. What I am expounding here is one modern view of the philosophy of science, that associated primarily with the name of Karl Popper.[2] Popperians would say that you could grade theories according to how easily they may be falsified; a theory incapable of being falsified or of being tested is not a good theory. You might as well throw it away.[3]

Let me restate, in colloquial language, what one might say from Popper's point of view about this beautiful experiment of Okabe and McNesby: We have, in the weakness of our minds, written down three and only three hypotheses for how ethane might fragment under ultraviolet irradiation. And in the strength and beauty of our hands and our minds, we have constructed experiments to eliminate two such hypotheses. That does not prove the third one at all. There may be a fourth or a fifth one we just were not clever enough to devise.

Now, everyone knows that. I know that, the people who did this experiment know that. But these are *people* who are doing experiments and interpreting them. It is in the nature of people *not* to want to write wishy-washy conclusions in papers, such as: "I have disproven A and B. I hope it's C, but maybe it's something else." No, people want to say, "I have proven C." Scientists want to do something positive.

30. The Salieri Syndrome

There is more. I now move from the specific and beautifully argued ethane photochemistry to a hypothetical but quite likely sequence of events. A piece of mechanistic chemical research gets published in a journal with a circulation of three thousand, goes to two thousand libraries around the world, a hundred people read the article, ten read it carefully. Among those ten is one very interested and interesting person, who all his life has been studying reactions of this type. As is in the nature of specialists, he has formulated very definite opinions on the mechanism of such reactions. And these young people, when they wrote their paper, did not even mention, did not even have a footnote to that older scientist's work! You can be sure that the slighted mechanistic chemist will be the world's most careful reader of this paper, and he will do absolutely everything in the world to try to prove that the authors' ideas are wrong. Is there something unethical about that, trying to prove other people wrong?

I do not think so, not at all. I return here to some statements about psychological motivations, already broached in chapter 18. Science is done by human beings. Human beings are motivated by a complex of things—among them curiosity and the search for knowledge. But there is also power, recognition, money, sex, and beauty, the same things that motivate other creators. Is there something wrong with this? Human beings are perforce fallible, yet capable of channeling their weakness into creation. There is nothing wrong with people for the "wrong reasons" thinking that some experiments might be wrong, and suggesting other mechanisms. As long as there are ten such people in the world, and a test that will confirm or deny an experiment, the system of science is OK; it will progress. But there is something in us that makes us think that it is wrong to do the right thing for the wrong reason. I am actually quoting from T. S. Eliot's *Murder in the Cathedral:*

> The last temptation is the greatest treason:
> To do the right deed for the wrong reason.[1]

Why do we think there is something wrong with the psychological imperatives of the dastardly slighted scientist in that little scenario I concocted? The reason is, I think, that we confuse the chemist's search for knowledge with a search for truth.

I think there is a potential danger in substituting truth for knowledge. By classifying ourselves as servants of truth, we place ourselves in the company of preachers and politicians. We should be, I think, with creative artists. First, because we do create this world. Second, because the public has fewer illusions about artists. We look for great art from good artists, but we do not look necessarily for better moral behavior than that of the average human. We want them to be moral and ethical, of course, but we know they are not angels. Why should we think scientists are?

You have read the revelations in the United States of various sexual misdemeanors on the part of some of our evangelical preachers. Why do we take such prurient interest in the various moral misdoings of the clergy? The reason is clear, of course. We know a priest is just human stuff—but we nevertheless confuse what he preaches with his personality. If he falls, he seems to fall further.

So it is in science. I suspect the interest in the few cases of fraud in science derives from similar causes, because we have built up our image, self-servingly so, as priests of truth.[2]

Let me call up a cultural referent, which is now, I think, well known to everyone—Peter Shaffer's play, *Amadeus*—either from a theatrical production or its film version.[3] The theme derives from a poem by Pushkin, "Mozart and Salieri."[4] You remember the story: Salieri cannot understand. At one point he says, in effect, "How could God have put such heavenly music into such a crass vessel?" We would like to think that Mozart was angelic, but in fact the great composer had a complicated personal and public life.[5]

Actually, most of the time we are willing to accept that artists may not be particularly nice people. Sometimes we even succumb to the equally romantic fallacy of attributing artists' creative impulse to their walking on the edge of sanity.

Since I alluded to Karl Popper, I do want to mention another pole of modern philosophical discourse in science. When I stress the human, often fallible psychological motives in science, I come close to the view of science held by Paul Feyerabend. Feyerabend was a polemical philosophical genius who consistently stresses that scientists are psychological and political beasts who will do just about anything to get their theories or their experiments accepted. Even though I read Feyerabend as inherently nihilist, inimical to science, what he has done is intriguing. He has shown in detail how scientists have selected data to prove their own theories. Here theorists are particularly susceptible (though experimentalists have other ways of deluding themselves). Feyerabend's *Against Method* is a good antidote to various romantic ideologies of the way science proceeds.[6] I think we need to recognize the strains of both Feyerabend and Popper in any real scientific activity.

31. STATIC/DYNAMIC

Let's return to look in some detail at what happens in that typical reaction flask. Or in the atmosphere, for that matter. In the process, another polarity inherent in chemistry will emerge.

You leave a glass of wine standing, and it evaporates. Wet clothes left on the line will dry. So you know something is happening. The molecules, natural or synthetic, you have come to believe in must leave the limpid liquid state and join their fellow travelers in the air.

Now let the wine (water *plus* alcohol *plus* a thousand—give or take a few—flavor ingredients) be sealed in its corked and leaded bottle. You know some vintages may last a century and bring a hefty price at an auction. Surely not much happens in the bottle. Oh, the wine changes, it may deteriorate, a deposit may form. But as it lies peacefully in the chateau bin it seems there can't be much molecular commerce between the liquid and the air trapped above it.

But there is. As there surely is in the deceptively quiet air of a peaceful room, in the bubble of old water trapped a million years ago in a rock, or across the membrane of a living cell, or even within a solid. In all of these "systems," as scientific jargon would have it, there is molecular motion unseen by the eye—seething, rapid movement in a gas, much slower motion in the solid. These are dynamic systems, only seemingly static. And that tension, of the only apparently quiet, is central to chemistry.

The sealed wine bottle reposing in the wine cellar seems quiescent for two reasons: The particles in motion are extremely small molecules. Even if they were slowed down they are too small to be seen with the naked eye or even an optical microscope. And their rapid (as it turns out) motions across the air-wine interface are precisely balanced in the sealed bottle—as many water (or alcohol) molecules in the liquid jump into the gas "phase" per second as different water (alcohol) molecules reenter the liquid. Overall it seems as if nothing is happening. The smallness of the actors, and the balanced nature of their actions, combine to make us think all is quiet on the molecular front.

The balance—as many in as out—is easy to get used to. Imagine a bathtub filled to a certain level. The plug is somehow pulled out, or maybe it never fit properly. The level of the water in the tub can remain constant if the faucet is turned on to just the right flow. The water in the tub changes—it is different water—but its level is the same. Two actions—water in, water out—are in dynamic equilibrium. The example seems wasteful, so you might think instead of that Milles fountain, where the water is recycled. Another analogy is to the number of people in a busy department store, with hordes entering and just as many leaving.

Note the contrast to static equilibrium, the (momentarily) unmoving scrum in a rugby match, the tightrope walker. These are tense states in and of themselves. The potential for a disruption of the equilibrium, the catastrophe of forces unbalanced, is too easily imagined. To be sure, the bathtub scenario has overtones of comic disaster as well—the plug falls in, the faucets cannot be shut off, there is no overflow outlet. Run for the mop. Where the heck is the main shutoff valve? This is an American bathroom, not a European one. So there is no floor drain. Call the plumber! (And maybe your lawyer!)

Dynamic equilibrium in chemistry can also be upset, sometimes with disastrous circumstances (a disease, an unwanted explosion). Most often, as we will see, we *do* want to perturb equilibrium, for our own purposes. But chemical dynamic equilibrium is not precarious. It is the stable state, the natural end. It even has restoring forces, resisting departure from equilibrium. These give it the semblance of a living thing and tempt us to use anthropomorphic language to describe a natural, inanimate balance.

31.1 A rugby scrum. (Photo by Robert E. Daemmrich/Tony Stone Images)

How did we learn that molecules in a gas and liquid are in rapid motion? People observed dust motes in a sunbeam, or the chaotic motion of smoke particles. Particles of such substantial size were moving rapidly, randomly. One could view them as buffeted by collisions with the invisible molecules of the air, known by the middle of the nineteenth century to be oxygen and nitrogen, and to be extremely small.

A theory was developed of the motions of such molecules in the air, subject to a few central assumptions:

- the pointlike molecules have all their mass concentrated in an infinitesimal volume
- they communicate with each other and the walls of their container only by collisions
- these collisions are *elastic,* a technical word describing that in their impacts only momentum is exchanged, that the molecules do not stick like putty or a thrown pie, but collide and bounce apart as steel ball bearings would.

The kinetic theory of gases, as this masterpiece of nineteenth-century physics is called,[1] gives us a description of the speeds and collisions of molecules. At this level of approximation (the molecules really are not pointlike objects, and they might stick a little to each other), the average speed emerges from the theory as a function only of the temperature and the mass of the molecules. Here is the formula for the average speed $\bar{s}$.

$$\bar{s} = \sqrt{\frac{8kT}{\pi m}}$$

T is the temperature in degrees C above absolute zero (T = °C + 273.15), m is the mass of the molecule, k a constant. The molecules are moving fast; table 2, opposite, lists values at room temperature and normal atmospheric pressure for a light molecule (H_2), a medium weight molecule (oxygen of the air), and a heavier one (diallyl disulfide, $CH_2CHCH_2SSCH_2CHCH_2$, one of the main constituents of garlic breath odor).[2]

31.2 One of the delights of life, garlic bread. Photographed by Joe Coca, from *The Garlic Book,* by Susan Belsinger and Carolyn Dille (Loveland, Colo.: Interweave Press, 1993).

Table 2

Some Predictions of the Kinetic Theory of Gases
(25°, one atmosphere pressure)

Molecule	Average Speed (meters per sec)	Average Distance Traveled Between Collisions (meters)	Average Number of Collisions per Second
H_2	1,770	1.24×10^{-7}	1.43×10^{10}
O_2	444	7.16×10^{-8}	6.20×10^{9}
Diallyl disulfide	208	1.42×10^{-8}	1.50×10^{10}

Note the enormous speed of these molecules; oxygen moves at close to the speed of sound (which is no accident—sound propagation depends on the molecular medium). The molecules don't get very far, however, before they collide with each other. The collision frequency and distance between collisions (called *mean free path*) do depend on the pressure and temperature of the gas. In outer space, the mean free path would be much, much greater (~10^9 kilometers in the intergalactic diffuse clouds; a friend remarks, "The poor chaps meet only every few hundred years").[3]

The average distance between O_2 molecules in our atmosphere is about 3.5×10^{-7} centimeters. This is about ten times the linear dimension of the molecule. One way to think about it is that the molecules through their rapid motions and collisions bang out an effective space around them that is substantially larger than the space they actually take up. It's a bizarre dance floor out there, in the seemingly quiet air.

So the theory goes, but do we know that molecules actually move at those speeds? Yes, we do. Here's an ingenious experiment by R. C. Miller and P. Kusch to probe not only the average speed but also the distribution of speeds (i.e., how many molecules move with any given velocity). In illustration 31.3 you see an oven at left, A. From a hole in it stream out many molecules of one kind. A beam is selected by a pinhole, B. The molecules at this point are flying through a vacuum, their collisions with each other being already done with when they emerged from the oven. They head for a notch helically carved in a solid cylindrical drum, C. The drum's speed of rotation may be varied. There is a detector, D, which measures the number of molecules emerging from the end of the helical notch.[4]

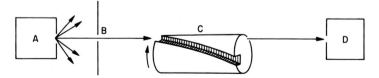

31.3 A schematic diagram of an experiment to measure the distribution of speeds in a gas.

You can see the devilishly clever setup. Only those molecules whose speed matches precisely the passage opening up ahead of them by the rotating drum, make it through. The others, too slow or too fast, just bang into the side of the helical groove. With a little algebra one can calculate from the "helicity" of the notch the speed of the molecules that make it through at a given rate of rotation of the drum. Then the rate is changed, allowing a batch with a different speed to make it through.

The outcome of Miller and Kusch's clever experiment matches perfectly the theoretical prediction (derived from the kinetic theory of gases in the nineteenth century) which is called the Maxwell-Boltzmann distribution. Illustration 31.4 shows the distribution of speeds in a component of the atmosphere, argon, at two different temperatures. Note that the average speed, close to the speed with which most molecules move (the top of the curve), is just one speed of many. Some molecules move slowly, some quickly.

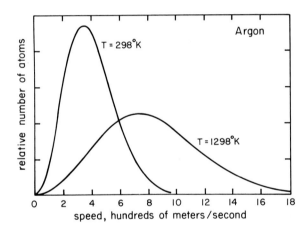

31.4 The distribution of speeds in a gas, argon, at two temperatures. Drawn after D. F. Eggers, Jr., N. W. Gregory, G. D. Halsey, Jr., and B. S. Rabinovitch, *Physical Chemistry* (New York: Wiley, 1964).

One piece of experience needs to be reconciled with the very rapid motion of molecules. Imagine the wearer of strong perfume entering a room. Or a skunk attacked by a dog across the yard. We know that released odors reach us much more slowly, on the time scale of seconds, than what we'd expect from the speed of sound, or the slower (but still very rapid) motion of the molecule(s) of perfume or skunk odor. Why is this so? The reason lies in those collisions—in air the perfume molecules do set off speedily in our direction, and surely that is the intent of the wearer of the perfume. But before those molecules get a fraction of a centimeter in your direction, they suffer many unplanned collisions with air molecules. The perfume eventually gets to us, but by a much slower random meandering called *diffusion*. Now in outer space, at least in that science fiction space opera, ah, there perfumed messages would reach their destinations so much more quickly . . .

32. Equilibrium, and Perturbing It

Let's move from that rapid motion of molecules to the idea of dynamic equilibrium. You and I need nitrogen for our proteins and nucleic acids. N_2 is 78 percent of the air we breathe, but we, that supposed pinnacle of evolution, don't know how to process N_2 biochemically. We get our nitrogen from plants, which in turn absorb it from the soil in the form of nitrate (NO_3^-) and ammonia (NH_3). But plants also don't know how to "fix" nitrogen from the atmosphere. Some bacteria, symbiotic with the roots of leguminous plants, do. The sources of plant nitrogen are: (1) to a small extent nitrates in the soil, (2) nitrogen fixed by lightning reacting N_2 with O_2 to eventually give NO_3^-, (3) the nitrogen fixed by bacteria, (4) manure, and (5) synthetic fertilizers. The man-made fertilizers (and modern machinery, farming methods, plant breeding, and pesticides) are responsible for the success of modern agriculture.

This is the background for writing down a chemical equation, for making ammonia.

$$N_2 + 3H_2 \rightarrow 2NH_3$$

It was recognized at the beginning of this century that one needed a steady supply of "fixed" nitrogen, and that ammonia would provide it. The reaction written above seemed the obvious way to do it; N_2 from the atmosphere is there for free, hydrogen gas is easy to make. If you took N_2 and H_2 and heated them you got some ammonia. But not much.

A German scientist looked at this problem and solved it in the period 1905–1910. The solution involves an appreciation of the dynamic nature of chemical equilibrium, and clever thinking on how to perturb that equilibrium.

Suppose you begin with N_2 and H_2 in a flask, and they react, to form NH_3. How do they do so? Not by action at a distance, but as a result of molecular collisions, perhaps a complicated set of intermediate molecules created, eventually leading to NH_3. The rapid motion of the molecules translates in a collision into the energy needed to break the strong bonds in N_2 and H_2. One must heat the mixture up to begin to break those bonds.

Suppose the reaction goes. Once some ammonia is created, it doesn't sit still. Ammonia molecules begin to collide with each other, and their energy makes them undergo the reverse reaction

$$2NH_3 \rightarrow N_2 + 3H_2.$$

The chemist summarizes the situation by arrows in both directions:

$$N_2 + 3H_2 \rightleftarrows 2NH_3.$$

Finally, equilibrium is reached—ammonia is formed by the forward reaction, decomposed by the back reaction. In that state of dynamic

32.1 Application of ammonia to a field (Farmland Industries, Inc.).

equilibrium the numbers of NH_3, N_2, H_2 molecules are *not* equal; but they are in some fixed ratio to each other. Everything seems to be standing stock still—nothing seems to move in terms of numbers. Yet underneath, as you've seen, there is tremendous motion.

But the outcome is an unhappy one, at least from the point of view of the selfish human being who wants to make ammonia, more ammonia, and nothing but ammonia. The dynamic chemical equilibrium system has restoring forces. There is no way of stopping the reverse reaction of NH_3 to N_2 and H_2, as much as we would like to do so. So what can we do?

The German scientist who looked at this problem knew that he had to define the condition of equilibrium, to *understand* it, before he went on to perturb it. He could have tried ad hoc moves, throwing at the system any catalyst off the shelf (this sometimes works, there's no denying serendipity). But it didn't work here. Understanding did.

How, then, to use equilibrium, apparently inimical to anthropocentric us? The chemist saw four strategies:

1. Take away the ammonia as it is formed. The equilibrium system, its restorative forces at work, will regenerate more ammonia.
2. Change the temperature at which the reaction is run. The reaction as written gives off heat. Lowering the temperature would, speaking loosely, absorb that heat, allowing more of the reaction to go left to right. More technically, the specific ratio of NH_3 to N_2 and H_2 at equilibrium is shifted by a decrease of temperature to favor NH_3.
3. Change the pressure. Note that the reaction takes four molecules (one of N_2, three of H_2), to two (of NH_3). There is a net reduction in the number of molecules. Since each molecule takes up about the same volume, the product (NH_3) side has a smaller volume. So if you increase the pressure in the reaction flask, the system will respond to this perturbation by producing more of the side which has a smaller volume (i.e., more NH_3).
4. Use a catalyst to help break the strong bonds in N_2 and H_2. This *is* an empirical business, prospecting for catalysts. The German scientist found, after much experimentation, that osmium or uranium were suitable catalysts.[1]

These strategies, making use of our understanding of equilibrium as a dynamic process, eventually worked. The industrial synthesis of ammonia, a workhorse of the chemical industry to this day, is the Haber-Bosch process.[2] In 1993 in the United States 3.45 x 10^{10} lbs of NH_3 were made this way. The process was devised by Fritz Haber, whose life, a cauldron of polarities, is the subject of the next chapter.

PART SIX

A Life in Chemistry

33. Fritz Haber

The creative chemist is moved by the problem at hand, and by general curiosity about the molecular world. Material support from society is certainly needed. For that support the chemist offers his or her energies in the advancement of reliable knowledge, once in a while even coming up with something practical. Who can blame him for wanting to be left alone most of the time; recalcitrant and beautiful matter presents enough problems of its own.

But this is not the way it is. The world has its ways of impinging on the creative scholarly life, of engulfing the person. The chemist would like the world to leave him alone, but it has its own ways of touching him, whether at the beginning, the middle, or the end of life. In no case that I know of has this fact been truer, or played out with more drama, than in the life of one of the greatest of all physical chemists, Fritz Haber.[1]

Haber was born in German Silesia in 1868, the son of a prosperous German Jewish merchant. Early in life he converted to Christianity, a fairly typical tactic of assimilated upwardly mobile Jews in Europe in the early part of the nineteenth century. By Haber's time, conversion was not necessary to achieve high status in the academic world (e.g., Richard Willstätter, one of this century's great organic chemists, did not feel the necessity for conversion). Nor did Albert Einstein. Haber did. And while he surrounded himself with Jews and people of Jewish ancestry throughout his life, he wore a convert's mask until close to the end of his life.

Haber's early years were marked by struggles with his father (his mother died only days after his birth). Interestingly, one of these difficulties involved a difference of opinion on the commercial role of synthetic dyes, the centerpiece of the then developing German fine chemical industry.

As much as Haber may have resented the early commercial exposure, perhaps it was the source of a unique talent that he showed later in life, for blending pure and applied science. One of his students, Karl Friedrich Bonhoeffer, later wrote of Haber:

> Free of all academic narrowness, he cherished in his work the close reciprocal relationship of technology and pure science. In this way he developed into a scientific personality whose intellectual concern was always devoted to preserving the ties between scientific progress and practical life.[2]

Haber did not have a great mentor. Nor did he begin his scientific career with a stellar success, a great synthesis, or the discovery of some great law of nature. Instead he labored largely by himself, on diverse problems in organic and physical chemistry. Throughout his life Haber had a tremendous capacity for work, and for assimilating the new. Fritz Stern, a thoughtful observer of the German historical and intellectual scene, makes the following point:

> From childhood on, Haber lived in historically dramatic times. His formative years coincided with the exaltation occasioned by Germany's unification, that belated achievement which gave the Reich its fatal militaristic-authoritarian character that even Bismarck at times regretted. . . . It would be foolish to draw too close a parallel between the development of the nation and young Haber, but the triumphs of both had something to do with feelings of inferiority which so many Germans wanted to exorcise. How many Germans transported their feelings of discontent of whatever origin in ceaseless work![3]

Haber's greatest achievement was the ammonia synthesis that I mentioned in the last chapter. It arose from a complete understanding of the constraints of chemical equilibrium, and what's interesting is that Haber was self-taught in physical chemistry. His eventual success also owed much to a determination, a stubbornness, that is perhaps exemplified by the following story, which according to Morris Goran, Haber told about himself:

> One very warm summer day he went hiking in the Swiss mountains. After a jaunt of eight hours, searching for drinking water, he came to a very small, seemingly uninhabited place. Water was not to be found, and he was very thirsty. Finally, he saw a well surrounded by a low wall. He immediately immersed his entire head. At almost the same time and unnoticed by him, a bull had done likewise; neither paid much attention to the other. But when they withdrew from the water, they found their heads had been interchanged. Fritz Haber had a bull's head and prospered as a professor from the eventful day.[4]

At the beginning of the ammonia story is a failure, and in its middle lies a scientific controversy, both of which only spurred Haber on.

Many had worked on the ammonia synthesis. In 1904 two Viennese entrepreneurs, the Margulies brothers, approached Haber to work on making ammonia from the elements. Haber and his students tried several metals, hoping to convert the N_2 to a metal nitride, which would then go on to react with H_2. But the temperatures needed were so high that little ammonia formed. The financial support from the sponsors dried up; the project seemed to be lost.

The failure rankled. Worse was to come in a questioning of Haber's data on the ammonia equilibrium by Walter Nernst, the dean of German thermodynamics. The point at question was the actual ratio of N_2, H_2, and NH_3 at equilibrium. Nernst had also worked on the ammonia synthesis at higher pressures. His theoretical understanding of what was necessary to achieve an effective synthesis was not inferior to Haber's. But Nernst had obtained a value of the "equilibrium constant" of the reaction

$$N_2 + 3H_2 \rightleftarrows 2NH_3$$

that indicated there would be less ammonia present at equilibrium than Haber had measured. Sufficiently less so that commercial synthesis was unlikely.

Haber and Nernst had clashed before, and were to do so again. In this case Haber took Nernst's experiments, done at higher pressures, as a challenge. Together with Robert Le Rossignol, he redid his experiments with great care and showed that Nernst was wrong.

More important, the controversy focused Haber's energies on the effect of pressure. Recall that the ammonia side of the equilibrium has two molecules, not four, as on the nitrogen plus hydrogen side. So an increase in pressure would favor the side of smaller volume (less molecules). This is the way to make more ammonia—except that the pressures required exceeded those used in the chemical reactors (glass and metal vessels) in use at the time. Haber and his coworkers, including a skilled metal worker, Friedrich Kirchenbauer, developed the containers and the methods of achieving the requisite high pressure, as well as the osmium and uranium (nothing to do with radioactivity) catalysts needed to help the reaction go at low temperatures.

Perhaps never before had a laboratory process for an industrial reaction been developed as thoroughly in an academic setting. Haber was fortunate in the sequel, in that the engineer who took over the process

33.1 Photograph of Fritz Haber, courtesy of the Eisner family. Hans Eisner was one of Haber's last students.

at BASF, then and now one of the world's great chemical companies, was the talented and ingenious Carl Bosch. Bosch developed a less expensive catalyst and transformed the reaction into an effective industrial synthesis. The Haber-Bosch process, perfected in small ways, is still in use today for the synthesis of most of those 3.45×10^{10} lbs. of NH_3 (see chapter 32).[5]

In my view, there is no doubt that Haber's achievement was and is a boon to humanity. The major use of ammonia is as a fertilizer (this in fact is the primary use of most of the chemicals produced in high volume in the world). This century has witnessed an incredible population explosion. Chemically intensive modern agriculture has managed to feed adequately (on the average, not without local famines) all those additional mouths. The yield from a good American acre of corn (150 bushels) is up by a factor of six since 1800. There is a case to be made for "organic" agriculture, but I think that synthetic fertilizers, and Haber's invention in particular, have prevented the starvation of hundreds of millions of human beings.

The Haber-Bosch process came on the scene just in time for Germany. With the outbreak of World War I in 1914, the German supply lines to South American fertilizer sources were cut. And most munitions contain much nitrogen, from TNT (trinitrotoluene) to ammonium nitrate (a fertilizer and an explosive used in the 1993 World Trade Center bombing in New York City). There were other industrial sources of nitrogen-containing compounds—coal distillation and the cyanamid process—but it can be argued that Haber's discovery was critical. A way of "making bread out of air" also proved to be essential to the war.

During the war Haber put the ingenuity and talent of his institute and his personal energies into the development of "chemical" weapons. (I put the descriptor in quotation marks to point out the absurdity of the differentiation—as if gunpowder, various metals, and explosives were not themselves chemical!) The Hague convention had outlawed "poison or poisoned weapons." There was some limited activity on both sides of the conflict prior to the war, but as L. F. Haber, who has written the definitive study of chemical warfare in World War I (and who is Haber's son), says:

> The most one can say about gas and smoke is that by the eve of the war military awareness of chemicals had increased to the extent that some soldiers were willing to consider them and a very few, with a more innovating turn of mind, were even experimenting with various compounds.

The substances used with the exception of phosgene, were not toxic. There were no military stocks of gases, nor of gas shell, save for very limited supplies of tear-gas grenades and cartridges in French hands. The forerunners were scientific curiosities and the belligerents of August 1914 had no conception of the practicalities of chemical warfare.[6]

They acquired them quickly. Haber's contribution was the concept of a gas cloud, his choice of chlorine and other chemicals, and his continued dedication. The German supreme command found in Haber "a brilliant mind and an extremely energetic organizer, determined, and possibly also unscrupulous."[7] He left decisions as to the legality of the use of poison gas to the high command.

Here is a description of the first large-scale gas attack, at Ypres, on the afternoon of April 22, 1915.

> The simultaneous opening of almost 6,000 cylinders which released 150 t of chlorine along 7000 m within about ten minutes was spectacular. The front lines were often very close, at one point only 50 m apart. The cloud advanced slowly, moving at about 0.5 m/sec (just over 1 mph). It was white at first, owing to the condensation of the moisture in the surrounding air and, as the volume increased, it turned yellow-green. The chlorine rose quickly to a height of 10–30 m because of the ground temperature, and while diffusion weakened the effectiveness by thinning out the gas it enhanced the physical and psychological shock. Within minutes the Franco-Algerian soldiers in the front and support lines were engulfed and choking. Those who were not suffocating from spasms broke and ran, but the gas followed. The front collapsed.[8]

Men had already died in so many ways in this war, as in other wars. But this was a new way. It was not an exclusively German way of killing, for the chemistry was in the end simple, and smart men and industry were there on both sides. Chlorine, phosgene, mustard gas, and chloropicrin were used extensively by Germany's opponents as well. Nor did poison gases just kill. Many more soldiers were injured, some badly; L. F. Haber estimates deaths as being 6.6 percent of all gas casualties.[9]

Rationalizers of gas warfare, then and now, ask, "Is there a nice, good way to die? What is worse about poison gas than shrapnel?" The answer is to be found in the testimony of the wounded. Something in the psyche, something deep that associates life with breath, is perturbed. Here is a section from Wilfred Owen's poem "Dulce et Decorum Est":

33.2 Training for chemical warfare. (Photograph by Jeffrey Zaruba/Tony Stone Images)

> Gas! Gas! Quick, boys!—An ecstasy of fumbling,
> Fitting the clumsy helmets just in time;
> But someone still was yelling out and stumbling
> And flound'ring like a man in fire or lime . . .
> Dim, through the misty panes and thick green light,
> As under a green sea, I saw him drowning.
>
> In all my dreams, before my helpless sight,
> He plunges at me, guttering, choking, drowning.
>
> If in some smothering dreams you too could pace
> Behind the wagon that we flung him in,
> And watch the white eyes writhing in his face,
> His hanging face, like a devil's sick of sin;
> If you could hear, at every jolt, the blood
> Come gargling from the froth-corrupted lungs,
> Obscene as cancer, bitter as the cud
> Of vile, incurable sores on innocent tongues,—
> My friend, you would not tell with such high zest
> To children ardent for some desperate glory,
> The old Lie: Dulce et decorum est
> Pro patria mori.[10]

The number of gas casualties among all combatants was relatively small (3 to 3.5 percent of all casualties by L. F. Haber's well-reasoned estimate).[11] The weather—wind, rain, heat—prevented then, and still does, the effective tactical use of chemical weapons in warfare. But the psychological stain of this weapon is indelible.[12]

I wonder if Haber, so experienced in catalysis, thought of poison gas (or of himself) as a catalyst, intended to speed up the outcome, to end the bloody stalemate of trench warfare. This was not to be. Germany lost the war. And another casualty was Haber's wife Clara, a chemist herself. She pleaded with her husband to give up his work on chemical weapons. He refused. We cannot know the causal connection, but she committed suicide.

After the war, Germany was saddled with a tremendous reparations debt of $33 billion, much of it payable in gold. Haber, the recipient (in 1918) of a Nobel Prize for the ammonia synthesis and now the leader of German chemistry, set his sights on extracting gold from sea water. He translated the total war debt into the equivalent of 50,000 tons of gold. The oceans were estimated by an Australian chemist, Archibald Liversidge, to contain 30 to 65 milligrams of gold per ton of sea water. This translated to 75 to 100 billion tons of gold in the oceans. The North Sea alone would do to the settle Germany's debt.

Haber did a series of experiments on "synthetic sea water," precipitating the gold ions with lead acetate and ammonium sulfide. He came to a conclusion that the gold could be economically separated if its abundance were even as low as 5 mg per ton of sea water. He then set about to check the previous literature estimates of gold concentrations, even equipping in high secrecy a Hamburg-American Line ship with a laboratory and extraction plant.

Haber was now in the business of analysis, which we have seen is an art and a science. Here is an account of what happened:

> Gradually however, problems arose. Haber covered vast areas of the Atlantic, and the water of Iceland and Greenland as well as the North Sea. He found the presence of gold varied considerably by region—for example, ten times as much gold appeared for a given volume in the North Atlantic as in the South Atlantic. Taking over 100 samples from offshore waters near the Californian gold fields, he found that even tidal changes made a great difference in results. Moreover, it appeared that when methods satisfactory for high concentrations were used with waters containing low concentrations, the results reflected the presence of gold in the reagents and vessels used. . . . Eventually, Haber decided that Liversidge was simply wrong; and on two points disagreed with him: gold nowhere exceeded .001 mg/m^3; and occurred with suspended matter rather than in solution.[13]

We come here to another tension characteristic of chemistry—suspicion and trust. Haber believed Liversidge's earlier analysis, as well as that of Edward Sonstadt, another chemist active in this field. In the papers he subsequently wrote,

> Haber divided his criticism between Sonstadt, who had undoubtedly been deceived by reagent contamination, and who in an article published in 1892, seemed to admit as much; and Liversidge, whom he faulted on technical grounds. Liversidge had used a method which required extremely sensitive methods of extraction. Sadly, in Haber's words "diese Vertrautheit hat Liversidge nicht besessen." Liversidge had simply produced results using unsatisfactory procedures.[14]

The modern alchemist was disappointed.

In early 1933 Hitler and the National Socialists came to power, with their baggage of anti-Semitic views. Already in April of that year they had issued a decree for purging Jews from the civil service. Haber's world broke apart; he, who was not really a Jew, now was a Jew. Haber had represented one pole of German Jewry—not only completely integrated into German culture but patriotic in the extreme. Albert Einstein represented another pole—German, but always suspicious of his native country. Haber, morally despondent, was crushed by the turn of events. Fritz Stern describes the situation:

> The silence of colleagues, the betrayal by the elites, was devastating. From exile, Einstein wrote Haber a letter full of compassion for his fate: "I can imagine your inner conflict. It is as if one must give up a theory which one has worked on all one's life. It is not the same for me because I never believed in it in the least." The theory was faith in German decency, in a future in which Jews and Christians could live and work together.[15]

Haber could have stayed on in his position since the law for the moment excluded war veterans from dismissal. He would have been forced, however, to dismiss his Jewish coworkers. Instead he resigned. This is an excerpt of his letter of resignation of April 30, 1933, to the Nazi Minister of Science, Art and Education:

> My decision to request my retirement follows from the contrast of the research tradition in which I have hitherto lived with the different views which you, Mr. Minister, and your Ministry advocate as protagonists of the present great national movement. In my scientific office my tradition demands that in choosing my collaborators I take into account only the

professional and personal qualifications of applicants without regard to their racial background. You will not expect a man in the sixty-fifth year of his life to alter a way of thinking which has guided him for the past thirty-nine years of his university life, and you will understand that the pride with which he has served his German homeland all his life now dictates this request for retirement.[16]

The minister said he was well rid of the Jew Haber. Now there was no mask; Haber wrote to Einstein in August 1933, "In my whole life I have never been so Jewish as now."[17] Fritz Haber left Germany for Switzerland, thought about taking a position in the country of his former enemies (England), thought about settling in Palestine. He was a broken man; this great German chemist died on January 29, 1934, in Basel, geographically close to his homeland but spiritually far removed indeed.

Less than ten years later, another product of the chemical industry, another gas, was used in the murder of millions of Haber's people in the extermination camps.

Part Seven

That Certain Magic

34. Catalyst!

Central to Haber's successful synthesis of ammonia was the invention of a catalyst for the reaction. Richard Zare, an insightful chemist, says, "If I had to pick any one word that to me most captures chemistry it would be the word *catalyst.*" No other field has its equivalent.[1]

Zare is right. A catalyst—something added to a reaction in small amounts that makes the reaction go faster, usually much faster, a substance that gets involved and yet is regenerated—is close to the heart of chemistry. Catalysis also touches two archetypal human themes:

1. Making easy what was perceived as nigh impossible, thereby overcoming an obstacle, and
2. The miracle of consumption and regeneration, of Persephone and the Resurrection.

Because these themes are part of our collective unconscious, the chemical catalyst is to nonchemists an object of fascination and yet, in some fundamental way, it is also *comprehensible,* something to grasp and identify with in a sometimes abstruse science.

Two examples help to show the generality of the idea: In Goethe's *Elective Affinities,* that unique novel of chemistry personified (and decades before the word *catalyst* was coined), there appears an odd character by the name of Mittler. He makes it "the central point of his life . . . never to enter any house where there was not a dispute to settle or difficulties to put right."[2] As his German name implies.

Almost two hundred years later, the American fashion designer Halston decides to introduce a new perfume. Their "nose" and market consultants create "Catalyst!" and splash it, replete with slinky erotic imagery, across several pages of the *New York Times Magazine*. The ad agency's copywriter must have had a chemistry course, for he or she writes in the ad:

> the climate is changing . . .
> disturb the equilibrium . . .
> Catalyst.
> For the woman of today,
> who has the power to change
> tomorrow.
> Feminine, romantic, a bit
> disruptive—Catalyst . . .[3]

The essential features of catalysis are easy to understand. Take a typical chemical reaction:

$$\text{Reactants} \qquad \text{Products}$$
$$A + B \quad \rightleftharpoons \quad C + D$$

We recognize that all such reactions are equilibria (i.e., that they occur in both directions). Nevertheless, very often (for every reaction we really want to *go*, or so it seems) we mix A and B, and very little happens. The reactants just sit there. In other words, and more precisely, equilibrium is not quickly established. Let's see why.

Reactants A and B are molecules made up of atoms bonded to each other in some special way. Products C and D are different molecules made up of the very same atoms. To get from A and B to C and D, chemical bonds must be broken and new ones formed. But it costs energy to loosen those old ties, and the advantages of the new bonding in the products may not yet be felt in the early stages of the reaction. The result—a barrier.

So we try a catalyst, a substance or compound, a molecule (often a mixture of molecules) that we add to the reactants. Call it X (indeed companies often hide the nature of their catalysts). X does not act by magic, at a distance. The catalyst X gets involved, initiating a sequence of reactions whose net result turns out to be the same as if the catalyst were not there. Here is the simplest example of a catalyst at work:

$$A + X \rightleftarrows AX$$
$$AX + B \rightleftarrows C + D + X$$

One or more of the reagents reacts with the catalyst, giving an "intermediate," a molecule AX. This intermediate is short-lived; it reacts quickly with another of the reactants, say B. In one step (or several), this second reaction generates the products C + D and reforms the catalyst X. Which is then ready to guide another set of reactants through the dance. Note the overall change is just

$$A + B \rightleftarrows C + D.$$

Lest you think this is just a formalism, let me immediately clothe this abstract scheme. By now, most everyone is aware of the problems that the thinning but essential ozone (O_3) layer in the atmosphere faces from chlorofluorocarbons. The ozone up there is formed and regenerated by certain natural processes. The chlorofluorocarbons are inert at sea level, but as they rise into the stratosphere they are decomposed by sunlight, generating chlorine atoms (Cl). The following sequence of reactions ensues:

$$Cl + O_3 \rightleftarrows OCl + O_2$$
$$OCl + O \rightleftarrows O_2 + Cl$$

The OCl is an "intermediate," a kind of counterpart to the catalyst, for this molecule is produced in the course of the reaction and then consumed. The oxygen atoms (O) which participate in the second reaction are not common at sea level, but are present in a sufficient amount thirty kilometers up to participate in the chemistry. The net reaction is

$$O_3 + O \rightleftarrows 2O_2$$

i.e., the conversion of an ozone molecule (O_3) and an oxygen atom (O) into two oxygen molecules (O_2). This is a process that goes on anyway; what the chlorofluorocarbons do is to provide another channel for depleting ozone. Not by magic, but through catalysis by chlorine atoms.

Note here another aspect of the same and not the same—the element oxygen appears in this reaction (and in nature) in three essential forms or *allotropes:* atomic O, molecular O_2 and O_3. Of these, the diatomic O_2 is the stable form under ambient terrestrial conditions.

Why does the reaction proceed quicker toward equilibrium in the presence of a catalyst? Because the energy barrier to the atoms of A and B rearranging to products is circumvented by the intervention of X. This catalyst molecule loosens the bonds (perhaps one at a time). It chops down that barrier. Not every X will do this, only certain ones, so there's room for ingenuity.

Catalysis fascinates because of its apparent magic:

(a) The catalyst X makes things happen that wouldn't have happened without it (not that we always want such changes; witness the ozone case).

(b) Very small amounts of a catalyst may transform a great amount of material. In principle, the set of reactions shown above could go on forever; in practice, the catalyst eventually gets depleted by some other chemistry.

34.1 Cover of February 17, 1992, issue of *Time*. © Time Inc. Reprinted by permission.

(c) X is regenerated. With that comes the snare for the observer, to believe that X is not involved. But it is; without its
chemical helping hands, nothing would happen.

The Cl from chlorofluorocarbons catalyzes an undesired reaction.
Next I want to show you two examples of helpful catalysts at work, both
in their own way essential to our way of life. One operates in the modern automobile, the other in your body.

35. Three Ways

No society is as dependent on the automobile—our servant, our master—as the United States. And none has had a more persistent romance with the car. The mass-produced, inexpensive automobile is essentially democratizing, the universal Volkswagen. Yet there is a litany of troubles that we may reasonably, I think, lay at the automobile's ever so appropriately thunking door. We have subsidized highways at the expense of more rational means of transportation, and have let our public transportation systems atrophy. We were blind so long to quality control and fuel efficiency that a very positive contributor to our trade balance lost its edge. The American automotive industry has come back, but late.

Is there any positive thing to be said about the United States and its passion for automobiles? Sure—through legislation and industrial ingenuity this country faced (and faces) the problems of automotive exhaust pollution before any other country did. A catalyst does it.

In the internal combustion engine, a hydrocarbon fuel—say, octane, C_8H_{18}—burns to give CO_2 and H_2O:

$$C_8H_{18} + 12.5O_2 \rightleftarrows 8CO_2 + 9H_2O$$

If this were all, there would be no problem. The CO_2 is a possible contributor to global warming, but not a major pollutant.

In fact, things are not ideal in that engine. First, a little hydrocarbon escapes burning and is just volatilized. Second, the burning may be incomplete, generating CO as well CO_2. Third, a small amount of the N_2, an innocent fellow traveler of the atmosphere's O_2 through the carburetor and into the combustion chamber, reacts at the high combustion temperatures with the O_2 (just as it does in lightning discharges). The products are a mixture of oxides of nitrogen, usually termed NO_x, prominent among them NO, nitric oxide.

All three of these side-products—hydrocarbons, CO, and NO_x—are pollutants. Under certain atmospheric conditions, in the presence of sunlight, elevated concentrations of hydrocarbons and NO_x may lead to photochemical smog formation. Our high-altitude benefactor, ozone, is formed, becoming a true villain at low elevations. The components of photochemical smog may cause eye irritation, impair breathing, and damage vegetation and materials. CO, the other pollutant, hinders motor skills at moderate levels by tying up some fraction of our hemoglobin, as I discussed in chapter 10.

The special problems of the Los Angeles basin have made the State of California a pioneer in leading the way for the nation in imposing exhaust emission controls. The automobile manufacturers screamed that they couldn't do it, but in fact, they came through. Pre-1966 automobiles were emitting 10.6 g of hydrocarbons, 84 g of CO, and 4.1 g of NO_x per mile driven. The 1993 California limits, met, are 0.25 g of hydrocarbons, 3.4 g CO, and 0.4 g NO_x. That is not a small decrease, but a factor of ten to forty.[1]

This great industrial achievement is mainly due to a catalyst called TWC, for *three-way catalyst* ("three-way" for its simultaneous handling of hydrocarbons, CO, and NO_x). The idea of the catalyst may be traced to a 30-year-old patent by G. P. Gross, W. F. Biller, D. F. Greene, and K. K. Kearby of Esso (now Exxon). The critical metal component in the effective catalyst, rhodium, was suggested by G. Meguerian, E. Hirschberg, and F. Rakovsky at Amoco in the 1970s. Catalytic treatment of automobile exhausts began in the United States with the 1975 model year; the first vehicles equipped with the TWC and needed electronic feedback systems were 1979 Volvos sold in California. The catalyst operates well only near a very specific air-to-fuel ratio of 14.65—therefore the need for associated precise fuel control.[2]

The TWC, like any good catalyst (or anything else that is the product of natural or human evolution) is a nice mishmash. Porous alumina (Al_2O_3) is coated on the channel walls of a ceramic honeycomb. In the alumina, or on its surface, are some other oxides: ceria (CeO_2), lanthana (La_2O_3), occasionally barium (BaO) or nickel oxide (NiO). About 1 to 2 percent of the materials applied to the alumina surface consist of the noble metals platinum (Pt), palladium (Pd), and rhodium (Rh), without which the catalyst is inactive. A cutaway view of a typical "catalytic converter" using the TWC is shown in illustration 35.1.

35.1 A "cutaway" of a typical catalytic converter. Photo courtesy of the Ford Research Laboratory, Ford Motor Company.

Some catalyst compositions contain all three metals, some don't. But none omits Rh, the most active of the three. A typical catalytic converter of a small car contains around a third of a gram of rhodium. Still other components may be added. Mordecai Shelef and G. W. Graham, two active researchers in the field, say: "This affords a very large number of permutations in deposition procedures and composition. As one would suppose, the exact 'Zusammensetzung' is the stock in trade of the catalyst maker, and is carefully guarded."[3]

The efficient rhodium is also very expensive—approximately three times the price of gold. Rhodium is obtained as a by-product of the winning of platinum; 74 percent of the world's supply comes from South Africa, 21 percent from Russia. In 1993, 90 percent of the world's Rh supply went into the TWCs.[4] It would be nice to replace Rh with another catalyst component, but nothing so far has been found to work as effectively.

How in fact does the catalyst work? I hope that you will not be disappointed if I tell you that there are pieces of information, partial knowledge, but as yet no certainty. Not for want of trying. The economic incentive to gain reliable knowledge of the innards of the process, and the likely power that would come with that knowledge to think rationally about replacing the Rh, is very great.

However, we can be certain that things do *not* happen in one fell swoop, with all the components—hydrocarbons, NO, CO—assembling neatly on the Rh surface, then rearranging into the products. The likelihood of that is really infinitesimal. The reaction is probably an extremely rapid sequence of simpler steps, involving one or two molecules at a time.

Here is one such sequence, concentrating on what happens to the NO and CO. It begins with the adsorption (a binding to the surface) of CO and NO:

$$NO(g) \rightleftarrows NO(a)$$
$$CO(g) \rightleftarrows CO(a)$$

Here (g) means "in the gas phase," and (a) means "adsorbed, bonded to the metal." A cartoon of what happens (only NO molecules are shown) is given in illustration 35.2:

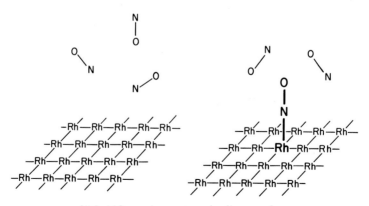

35.2 NO coming onto a rhodium surface.

There is excellent experimental evidence for such "chemisorption" of NO. Next, some people think that two NO's on the surface get together, thus forming an N-N bond:

$$2NO(a) \rightleftarrows (NO)_2(a)$$

This is actually a step that I and Tom Ward have worked on, from the theoretical side.[5] We postulate a reaction path as shown in illustration 35.3 and have figured out why Rh works better than Pd or Pt.

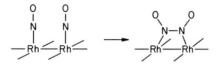

35.3 A hypothetical reaction path for coupling two chemisorbed NOs.

Remember, I said "some people." Not all. Where there is partial knowledge, there is controversy. Some people favor another sequence of events. The experiments needed to distinguish between the various mechanisms are not yet doable. The controversy remains.

Next, it is thought that a nitrous oxide (N_2O) molecule is formed and trapped on the surface:

$$(NO)_2(a) \rightleftarrows N_2O(g) + O(a)$$
$$N_2O(g) \rightleftarrows N_2O(a)$$

The nitrous oxide then falls apart on the surface, releasing nice, harmless N_2:

$$N_2O(a) \rightleftarrows N_2(g) + O(a)$$

Note that several of these steps have formed oxygen atoms (not molecules) chemisorbed on the surface. These then complete the "burning" of CO (and, in another reaction, of the hydrocarbons):

$$CO(a) + O(a) \rightleftarrows CO_2(g)$$

Somehow I doubt that it is as simple as this.

But the catalyst works. I don't want us to lose sight of the incredible Edisonian achievement of the TWC makers—a reduction of automotive emission pollutants to a few percent of what they were thirty years ago. We do not yet understand how these catalysts work. This is our weakness (of discovery), a counterpoint to our ingenuity (in creating them). It is also our challenge.

One final point. Recall from chapter 10 the deadly mimicry of carbon monoxide. A careful analysis by Mordecai Shelef concludes:

> In 1970, before the implementation of strict controls on emission in motor vehicle exhaust gas (MVEG), the annual USA incidence of fatal accidents by carbon monoxide in the MVEG was 800 and that of suicides 2000 (somewhat less than 10% of total suicides). In 1987 there were 400 total fatal accidents and 2700 suicides by MVEG. Accounting for the growth in population and vehicle registration, the yearly lives saved in accidents by MVEG were 1200 in 1987 and avoided suicides 1400.[6]

The findings here parallel a 35 percent reduction in the total suicide rates in England and Wales in the 1960s, traced to a single cause—the reduction in CO concentration in domestic gas. Gas used to be made from coal and contained up to 14 percent CO. The switch was then made to North Sea natural gas, which has very little CO.[7]

36. CARBOXYPEPTIDASE

The three-way catalyst was a mix of metal particles—platinum, palladium, rhodium. Atoms of another metal, zinc, figure in the action of the biological catalyst, an enzyme, that I will describe next. Enzymes are proteins, chains of amino acids. They are almost entirely made up of C, H, O, N, and S atoms. But often the active site of an enzyme uses essential metal atoms—among others, iron, copper, manganese, molybdenum, magnesium, zinc. Their importance in biological systems does not correlate with their abundance in the crust of the earth.

Carboxypeptidase A is a digestive enzyme. The molecules animals ingest need to be broken down before they are reformed into bigger and better molecules. Not all the way to atoms, no—that would be inefficient. Building blocks of two to eleven carbons will suffice, a set of molecular building blocks from which the diversity of biochemistry may be fashioned.

We eat (we need) proteins. Carboxypeptidase A is a *protease,* an enzyme that chops up protein by unhooking an amino acid from one end of the polypeptide chain. It specializes, as proteins do, in certain amino acids. Here is its chemistry (illustration 36.1):[1]

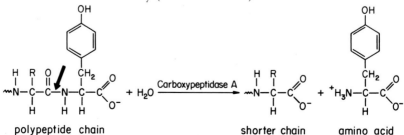

36.1 The chemistry accomplished by carboxypeptidase A.

Note the simplicity of the reaction; water is just added across a C-N bond marked by an arrow at left. So who needs the enzyme? We do. This simple reaction, $A + H_2O \rightleftarrows B + C$, does not go at a useful rate in the absence of the enzyme.

Let me be more specific: for a certain peptide Daniel Kahne and W. Clark Still measured that half of the peptide's C-N bonds would be broken in seven years (without an enzyme). That's a long time to wait to digest a hamburger![2]

In biochemical systems, the substrate (the molecule the enzyme acts on, typically denoted by S) and the water are delivered to the enzyme (E). The latter does its apparent magic by a sequence of reactions, then releases the products (P) and repeats its work. We can summarize the process schematically by the reaction sequence

$$E + S \rightleftarrows ES$$
$$ES + H_2O \rightleftarrows E + P$$

Here ES is a "complex" of the enzyme with the protein to be degraded, the intermediate.

We are curious, we want to know how the enzyme works. To find that out, we certainly want to begin with the enzyme's structure. But we also want to know the shape of the intermediate ES. And . . . that is like catching the wind. The enzyme is not called an enzyme for nothing—it catalyzes the relevant chemistry most efficiently. ES is there, but ever so fleetingly; the enzyme factory typically processes 100 million molecules a second.

The trick then is to slow down the process. By experimentation you can find an S' that binds to the enzyme, but is *not* chopped up expeditiously. ES' then sticks around long enough for us to do our structure determination. Chemistry is continuous and consistent. S' is the same and not the same as S. What we learn from the structure of ES' (there long enough for study) applies, probably, to ES (here and gone in a wink).

For carboxypeptidase A the structure of the isolated enzyme and several enzyme-substrate complexes were established by William N. Lipscomb, Jr. and his coworkers.[3] Lipscomb happens to be one of my two Ph.D. mentors, though I worked on chemistry far away from the enzyme.

Carboxypeptidase A consists of a single polypeptide chain about 900 atoms long, with many attached groups, 307 amino acids in all. It folds up into a compact shape about 50 x 42 x 38 Å (1 Å = 1 x 10^{-8} centimeter; for calibration, an oxygen molecule is about 3 Å long). Illustration

36.2 shows a piece of its structure, and that of the complex (ES') with a slowly cleaved substrate, glycyltyrosine:

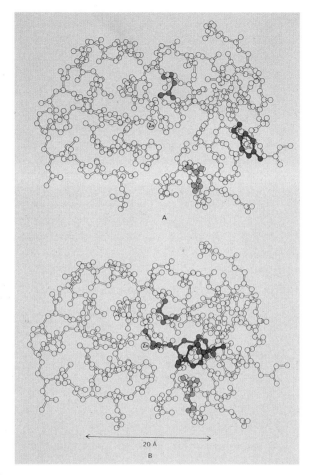

36.2 The structure of carboxypeptidase A (E, *top*) and its complex with glycyltyrosine (ES', *bottom*). After W. N. Lipscomb, *Proceedings of the Robert A. Welch Conference on Chemical Research* 15 (1971): 140. Actual illustration is from L. Stryer, *Biochemistry*, 3d ed. Copyright © 1988 by Lubert Stryer. Used with permission of W. H. Freeman and Co.

Illustration 36.3 enlarges a piece of the enzyme-substrate complex (ES', initially shown in illustration 36.2, right); the bound glycyltyrosine is in red.

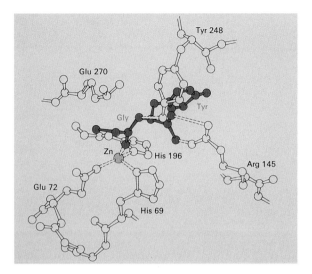

36.3 The environment of the glycyltyrosine bound to carboxypeptidase A. After D. M. Blow and A. Steitz, "X-ray Diffraction Studies of Enzymes," *Annual Reviews of Biochemistry* 39, no. 79 (1970). Copyright © 1970 by *Annual Reviews Inc.* All rights reserved. Illustration is from L. Stryer, *Biochemistry,* 3d ed. Copyright © 1988 by Lubert Stryer. Used with permission of W. H. Freeman and Co.

There is a place for the substrate to be sure—a groove, a cavity, a keyhole, whatever metaphor of relative location you choose. But things are a little more complicated. The groove, the cavity, the lock into which the key fits, is not static. It breathes. Or, to be less anthropomorphic about it, the enzyme readjusts its shape as it binds the substrate. That this happens often has been argued cogently by Daniel Koshland, Jr., in his "induced-fit model" of enzyme action.[4] It may be seen that a couple of atoms of one amino acid (tyrosine 248, marked in blue) move as much as 12 Å (one quarter of a linear dimension of the enzyme), in response to the binding.

Here is what Lubert Stryer says of the molecule:

The bound substrate is surrounded on all sides by catalytic groups of the enzyme. This arrangement promotes catalysis. . . . It is evident that a substrate could not enter such an array of catalytic groups (nor could a product leave) unless the enzyme were flexible. *A flexible protein provides a much larger repertoire of potentially catalytic conformations than does a rigid one.*[5]

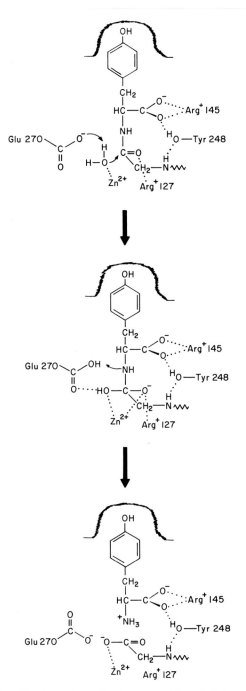

36.4 A mechanism for the action of carboxypeptidase A. Adapted with permission from D. W. Christianson and W. N. Lipscomb, *Accounts of Chemical Research* 22 (1989): 62–69. Copyright © 1989 American Chemical Society.

In the end Lipscomb's beautiful work has led to a detailed mechanism for carboxypeptidase A's magic cleaving. This is shown in illustration 36.4. The bound enzyme complex (ES) is attacked (top of illustration) by a water molecule that is "activated" by a zinc ion and a specific amino acid of the protein, glutamate 270 (Glu 270; the numbers here refer to a serial ordering of the amino acids). Another intermediate complex forms (middle of illustration); as it collapses, the C-N bond is cleaved and a hydrogen added to the nitrogen again from that critical Glu 270. Not without a little help from my friends—enzyme amino acids Arg 145, Try 248, and Arg 127: the bottom panel of the illustration shows the pieces just before they are released.

Nature offers no more apologies for the beautiful *complexity* of this process than it does for the multitude of life in a square meter of your backyard or the U.S. Supreme Court's considerations of abortion rights.

Mircea Eliade, a historian of religion, has written a remarkable book, *The Forge and the Crucible,* which traces the relationship between religion, metallurgy, and alchemy. In his beautiful concluding chapter, Eliade makes the haunting observation that the goal of the alchemist was to hasten the "natural" evolution of metals from base to noble, and to secure a similar transformation of the body, from sick to healthy, from mortal to eternal. The alchemists failed, in the end, and were replaced by modern chemists and physicians, who, denying a connection all the way, have achieved, through catalysts, composites, and pharmaceuticals, a very large part of the alchemist's original goal.[6]

Part Eight

Value, Harm, and Democracy

37. Tyrian Purple, Woad, and Indigo

The duality of benefit and potential harm is faced by real, fallible, ethical human beings in the context of any object in their environment. An automobile, a bread knife, a television program may help or hurt. But this duality comes to the fore in contemporary attitudes toward the workings of the great chemical industries of this world. When considering the wages, the bountiful products, the wastes of our giant and small factories, you are truly contemplating the Janus image.

There always has been a chemical industry, for it is impossible to live in this world without transforming it. Many practical protochemistries—metallurgy, cosmetics, fermentation and distillation, dyeing, apothecary formulations, and the preparation of food—were with us for thousands of years before the science of molecules. Early on, the objects of these transformations became the stuff of organized commerce.

I think, for instance, of the marvelous elite manufacture of the pigment called Tyrian purple.[1] From early on in the history of Rome (and in the lives of the Hebrews) a purple dyed wool, of a hue ranging from red to blue-black, was highly valued. It was called Tyrian purple or royal purple. Pliny the Elder described it as "the color of congealed blood, blackish at first glance, but gleaming when held up to the light." In Republican Rome clothing completely dyed in purple could be worn only by the censors and triumphant generals, while the consuls and praetors wore purple-edged togas, and generals in the field a purple cloak.

The manufacture of royal or Tyrian purple was highly restricted in the Roman Empire. It became a capital offense to manufacture the royal purple other than in the imperial dye-works. Meanwhile, the Hebrews wrote a prescription for a blue into the Old Testament, specifying that one strand in the fringes of their garments had to be dyed a specific blue, called *tekhelet*.

The Tyrian purple and the biblical blue are of animal origin. They were extracted painstakingly, and therefore expensively, from three species of gastropod snails: *Trunculariopsis trunculus, Murex brandaris,* and *Thais haemastoma.* Illustration 37.1 shows the shells of these three species. In one of the body structures—the mantle—of these beautiful shelled snails, there is a hypobranchial gland. This chemical factory has numerous functions and produces not only a mucoid substance that cements particles as they are expelled by the snail but also several neurotoxic chemicals used in predation. And it emits a clear fluid, which is the precursor of the dyes. On exposure to the oxygen of the atmosphere, under the action of enzymes, and importantly, sunlight, the fluid changes from whitish to puslike yellow, then green, and finally blue and purple. Aristotle, a careful observer, and Pliny both give good descriptions of the snails and the process of dye extraction.

37.1   The three snail species yielding Tyrian purple (*from left to right: Murex brandaris, Trunculariopsis trunculus,* and *Thais haemastoma*). Photo by D. Darom. Reproduced by permission from E. Spanier, *The Royal Purple and the Biblical Blue* (Jerusalem: Keter, 1987).

The snails had to be correctly identified, their shells carefully broken, the precious mantle fluid collected and allowed to react, the dye separated, concentrated, and the wool or silk prepared for dyeing. There may have been a simple chemical procedure, a reduction-oxidation sequence. This was needed to make the dye soluble, then to fix it in the fiber. We have archaeological evidence of this ingenious chemical activity along the eastern shore of the Mediterranean. It seems the Phoenician chemists had waste-disposal problems; their shell dumps are extant.

All along, there has been another, much more economical, source of a dye very closely related to the royal purple and the biblical blue. It is from the genus *Indigofera* of herbs of the pea family, widely dispersed in warm climates. This plant was an important product of the India trade, for it is readily cultivated there. A field of indigo appears

37.2 (*Top*) A field of indigo in California, photographed by D. Miller. (*Bottom*) A field of woad in France, photographed by W. Rauh.

37.3 The production of indigo, from the *Encyclopédie* of Diderot and D'Alembert.

in illustration 37.2, top.[2] And from Diderot's classic encyclopedia of 1753, illustration 37.3 shows the production of indigo dye. Fermentation and oxygenation stages took place in these vats.

Another source of the purple dye is the woad plant, *Isatis tinctoria* (illustration 37.2, bottom).[3] It is widespread in Europe and across to Asia. The plant was widely used in more northerly climates until it was displaced by the southern indigo plant of the East India trade.

And what is a pea plant doing making a molecule identical to that synthesized by a snail? A good question. Surely it has to do with the common biochemical pathways shared by living organisms, and the wondrous games evolution plays. For other examples of species of very different phyla making the same complex molecules, I would point to nepetalactone, the active principle of catnip, which comes from the mint as well as from a walking stick (an insect), and the cardiotonic bufadienolides, from the venom of toads and from fireflies.[4]

In the second half of the nineteenth century we learned that the purple color from snails, the indigo plant, or woad was due to a molecule, named indigo and having the structure shown in illustration 37.4. From some animal species one also finds a related molecule in which two hydrogens are substituted by bromines.

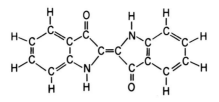

37.4 The chemical structure of indigo.

And then, in the last quarter of the nineteenth century, with the science of chemistry in ascendance, German chemists learned how to *synthesize* indigo. While they may have developed the synthesis in part for curiosity's sake, their aim was clearly utilitarian and commercial— there was a market for dyes, and for this one in particular.

38. Chemistry and Industry

What transpired between the mollusk fisheries and Tyrian protofactories and the successful mass production of synthetic indigo around 1900 by the Baeyer, Degussa, and Hoechst companies? Quite a lot. The scale of transformation of the natural took a great leap. The Tyrian purple protochemistry took a natural product and without much understanding but with great care and skill (does that sound familiar?) transformed it into a product of utility and desire, therefore of commercial value. The German dyestuffs industry also started with natural raw materials—first coal tar, then petroleum, and ethanol, potash, acetic acid as well. But the nineteenth-century industrial synthesis involved many stages. A chemical process grew into what we know today, a sequence of hundreds of physical operations, carried out in gleaming glass or steel vessels—an operation large enough to produce synthetic indigo to dye x million pairs of blue jeans every year.[1]

38.1 Some early samples of synthetic dyes from the laboratories of BASF. From the collection of the Deutsches Museum, München. Photo courtesy of Otto Krätz.

In the second half of the nineteenth century the German dye industry grew dramatically, diversifying to chemotherapy, fertilizers, and explosives. There is nothing specifically German here; the knowledge,
like all chemical knowledge, was and is universal. A larger and larger
part of the gross national product of all industrialized countries became chemical in nature.

Directly or indirectly, the wealth of nations depends on chemistry—
on their collective capability to transform the natural. In defining the
role of chemistry in the world economy I would include all transformations of natural matter, including food processing, the winning of metals (a very chemical process), and the production of energy (could you
call the combustion of oil, coal, or natural gas anything but chemistry?). By my reckoning, chemistry would then play a major role in
nearly a quarter of the GDP of an industrialized country. Most econo-
mists limit the definition to the chemical process industries—still
broad enough to include synthetic fibers and plastics, bulk chemicals,
fertilizers, fuels and lubricants, catalysts, adsorbents, ceramics, propel-
lants, explosives, paints and coatings, elastomers, agricultural chemicals, and pharmaceuticals. And more. The U.S. chemical process industry sold 4.32×10^{11} worth of goods in 1990, adding more in value to
the raw materials than the cost of those materials.

38.2 A petrochemical plant in Holland, photographed by Ian Murphy/TSI.

In the United States the chemical industry is one of the few large components of the economy which contributes positively to our net balance of trade, which as we all know is net negative. Illustration 38.3 shows several components of that balance—note that the only positive bright lights are for chemicals and aircraft.[2]

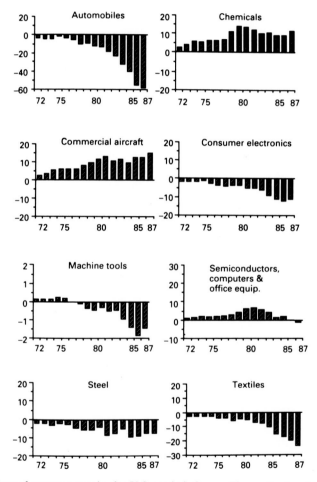

38.3 Several components in the U.S. trade balance. The vertical scales are in billions of dollars.

Table 3 lists the top twenty hits of the chemical world in 1993.

Table 3
Top Twenty Chemicals in the United States

Rank	Chemical	1993 U.S. Production (in billions of lbs.)
1	Sulfuric acid	80.31
2	Nitrogen	65.29
3	Oxygen	46.52
4	Ethylene	41.25
5	Lime (calcium oxide)	36.80
6	Ammonia	34.50
7	Sodium hydroxide	25.71
8	Chlorine	24.06
9	Methyl *tert*-butyl ether	24.05
10	Phosphoric acid	23.04
11	Propylene	22.40
12	Sodium carbonate	19.80
13	Ethylene dichloride	17.95
14	Nitric acid	17.07
15	Ammonium nitrate	16.79
16	Urea	15.66
17	Vinyl chloride	13.75
18	Benzene	12.32
19	Ethylbenzene	11.76
20	Carbon dioxide	10.69

SOURCE: Data from "Facts and Figures for the Chemical Industry," *Chemical and Engineering News* (July 4, 1994): 31.

I assure you that these chemicals are not made in such quantities for fun. Someone buys them, someone uses them. And not just for luxuries, but for bread, both literally (because agricultural fertilizers, for instance, are the main fate of the number one hit, sulfuric acid) and figuratively speaking. But the production of these vast quantities of chemicals does occasionally cause problems.

It's interesting to think about the properties and end uses of the top twenty. Chemistry students spend a lot of their time studying acids and bases. For good reason. Among the top twenty are three acids (sulfuric, phosphoric, and nitric) and three bases (lime, sodium hydroxide, and ammonia). Acids and bases are the initiators of change. They react.

Modern agriculture has fed adequately, not perfectly, an incredibly rapidly growing global population. The primary responsibility for this success lies in the utilization of chemical fertilizers. There *are* some problems of modern chemically intensive agriculture: fertilizer runoff directly affecting water life, wastes produced in the manufacture of ag-

ricultural chemicals, damage to us and other living creatures from herbicides and pesticides, interference in the grand cycles of the earth,
global climate change. These problems are real. But the hungry
mouths of children around the world cry out for food—and chemical
fertilizers made from no less than seven of the top twenty molecules
help us to answer that cry.

The top twenty list changes slowly. In 1993 no new chemical entered
the list, none dropped out of it. But over a longer time interval—say,
fifty years—there is change. The new kids on the block since 1940 are
ethylene, methyl *tert*-butyl ether, propylene, ethylene dichloride, vinyl
chloride, and ethylbenzene. All but one of these are the raw materials
of the polymer century, the sources of plastics and synthetic fibers.

Incidentally, were gasoline considered officially as a chemical, it
would be number one on the top hit list. In the United States about
six times as much of it flows into the gas tanks of our cars as sulfuric
acid is made. The amount of gasoline consumed is so vast that even
though the fuel itself is not on the list, a gas additive is. This is the
rising star, methyl *tert*-butyl ether (illustration 38.4), sometimes abbreviated as MTBE. It jumped 121 percent in its production over the year
before (so maybe you should have invested in the companies making
it)! MTBE's presence in the top twenty, its spectacular ascent, are testimony to our infatuation with the automobile and to the way science,
technology, environmental concerns, and government regulation interact in this world. Methyl *tert*-butyl ether is the viable alternative to
hazardous tetraethyl lead as a gasoline additive, raising the "octane"
rating. It may be present to the extent of 7 percent in every liter of gasoline.

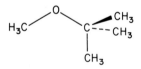

38.4 The chemical structure of methyl *tert*-butyl ether, the rising star.

What will be the next molecule to make the charts?

39. ATHENS

Something else of considerable note happened between the utilization of Tyrian purple indigo protochemistry and our present time, something in the larger world itself. An old idea, democracy, grew into the souls of people. The notion was that men (and God knows it took 2,400 years to see that women had that prerogative too) had the right to govern themselves. The idea was that the social contract implied a given equality at the beginning, so that if men and women lived together, that the legitimacy of their actions (delegated in some way if need be) stemmed ultimately from themselves and not from a master or king or tsar or party secretary or ayatollah.

It is worthwhile to reflect on democracy in this time, some 2,500 years after Cleisthenes' reforms in Athens, and a few decades after democracy's return to that beautiful and ancient land of Greece. That this idea had to return, and not only once, reflects that ancient struggle between forms of government, between democracy, oligarchy, and tyranny. That contention continues, and has meaning in our time. I want to look at this social invention, democracy—as much the work of men and women as the 10^{11} pounds of sulfuric acid made every year—in the context of its interaction with science and technology.

Chapters 39 and 42 are adapted from my Seferis Prize Lecture of the Fulbright Foundation, published in *Khimika Khronika* 54, no. 1 (1992): 4–8.

39.1 An Athenian vase (a *dinos*) by Sophilos, illustrating the Wedding of Peleus (British Museum, London). Note the figure of the centaur Cheiron at upper left—he will recur in our story. Classical Greek vases were colored by compounds of iron and manganese; the firing conditions were important in determining the color.

The basic content of Classical Athenian democracy is clear, even as its radicalism varied. The polity granted the right to be heard, and a voice in decisions to all citizens. True, women, slaves and that interesting category of resident aliens, the *metics,* were excluded. But we must not ask too much, to hold the past up to today's standards. The city-state also demanded in return service, to a degree unparalleled since. Much of that service was in the political sphere. The Athenian democracy was participatory, in the oral or spoken sense, and all-embracing of its citizens. Imagine that in a city of 17,000 citizens, a jury votes on the guilt of Socrates by a vote of 280 to 220! And it wasn't the only jury, the only *dikasteria* proceeding that day! Nine others might have been underway at the same time.[1]

Trust in the people, a separation of the public (*to koinon*) and private (*to idion*) spheres, a social contract between the individual and the state, all these are the lasting contributions of Greek democracy. That it, in its Classical Athenian form, did not survive, is only testimony to the eternal struggle for justice and basic human rights. I remind you how that struggle continues, this day I write, in Burma, in Cuba, in Iraq, in those remarkable events we have seen with our own eyes in Eastern Europe. And neither we, nor the Chinese people, will forget the early days of June 1989 in Tiananmen Square.

40. The Democratizing Nature of Chemistry

Science and technology have transformed this world, mostly for the better (but with some ill consequences). The claim I want to make here is that the effects of science, and chemistry in particular, are inevitably democratizing.

The world that my great-grandparents were born into 150 years ago in the Austro-Hungarian province of Galicia, or the world of the backwaters of Zaire today, was not a romantic paradise. The world was—and for many people living today remains—a brutish, inimical environment. Perhaps one lived in balance with it, but with a life span far from biblical. One only has to look at the cemeteries of the last century, or read the heart-breaking diaries of our ancestors, to see the tragedy of seven children out of eleven dead before puberty, or of childbirth as a killing prospect. When I hear an opponent of technology speak against modern, chemically intensive agriculture or pharmaceutical therapy, my heart beats quicker in a rush of anger at the implicit lack of simple human sympathy in his stance.

Humankind has seen a doubling of our life span; less death and suffering; birth control; a greater color palette to lift the spirit; freedom from the smell of sewage; a way to cure many (though hardly all) diseases; more light and food for all, and generally better air quality; and food for the soul in the Ramayana on the screen or a Mozart rondo in the air—these are things of which scientists and engineers really can be proud.

Technology and science also serve humanity's evil side, as aspects of subjugation, propaganda, and even torture. Some would see in this the ethical neutrality of science, and even a reason to condemn it. Quite aside from the misuses of science, to many in the low-income economies science may appear as a luxury of the elite, or just another element by which the privileged classes oppress poor people.

Simple technological solutions centered on improving the human position also have ways of eliciting "countermeasures" from nature. No need for confrontational language here; it's just a complex, interrelating system, one that has *evolved,* reacting to change. The same chemically intensive agriculture and antibiotic therapy that has made life better also elicits the natural selection of pesticide- and drug-resistant forms of life. But I really do think that the overall effect of science is inexorably democratizing, in the deepest sense of the word—by making available to a wider range of people the necessities and comforts that in a previous age were reserved for a privileged elite.

41. ENVIRONMENTAL CONCERNS

Political democracy is a social transformation as irreversible as chemistry, the science of the transformation of matter. I need to mention this because I perceive in the attitudes within my own profession today some strands of thought which seem to me to be forgetful or skeptical of the process of democratic governance.

Let me caricature some prevailing attitudes in the chemical profession. We say that we're reasonably well off in the material reality of this world, in our remuneration (but never rewarded sufficiently, of course), in what we really contribute to society. But spiritually it's a different story. We ain't got no R-E-S-P-E-C-T. We're typed by society, so the complaint goes, as the producers of the unnatural, and collectively labeled as polluters. We are surrounded by "chemophobia," by an unreasonable, irrational fear of what we do. The media seem to be engaged in a conspiracy against us, and what expertise does Meryl Streep, the great American actress, have to testify to Congress about what's in our apples? In fact, let me use that Alar story—which is where Meryl Streep comes in—to make some points about chemistry and democracy.[1]

Chapters 41, 44, and 45 are adapted from my Priestley Lecture for the American Chemical Society, published in *Chemical and Engineering News* 68, no. 17 (April 23, 1990): 25–29.

Alar, or daminozide, a growth regulator, is one of perhaps two dozen chemicals that may be legally applied to apples during their maturation process. It keeps the apples longer on the tree and helps the maturation of firmer, more perfect fruit. A very small fraction of Alar is absorbed into apples and metabolized to an unsymmetrical dimethyl hydrazine—or UDMH, for short. The levels of UDMH in apples are probably insufficient to have biological effects on humans. A public awareness group, the National Resources Defense Council, brought out the use of Alar and in various alarmist ways publicized the carcinogenicity of the UDMH metabolite. Alar-treated apples, already of some concern (reasonable or not) to supermarkets selling them, were quickly pulled off the shelves. Eventually, Uniroyal Chemical, the producer of Alar, halted sales of the hormone.

41.1 Apples on a tree in the Cornell orchards.
(Photograph by Jay A. Schwarcz)

Many chemists reacted to this episode instinctively by (1) "tut-tut"-ing the concerns, (2) impugning the motives of the public awareness group and Ms. Streep, and (3) pointing to this story as a typical, irrational example of chemophobia.

That wasn't my reaction. I must admit, however, that I wasn't consistent and tended to fall into the three stances I just enumerated some of the time. But my initial reaction as a chemist and a human being was, "Gee, I didn't know there were synthetic chemicals in my apples!" I didn't know Alar existed. To be sure, I knew apples were treated in various ways, with fertilizers, herbicides, insecticides, fungicides, ripening agents. I had been trained since childhood to wash off fruit, for getting dirt off it. Subtly over the years, the real reason for washing fruit changed to removing any chemical residues. (Am I the only one to have this feeling? I don't think so.) But I didn't know, or maybe I didn't want to know, what might have found its way inside, what had not been degraded. I didn't know what remained inside, such as UDMH, at what levels, and what were its biological effects. I didn't like that—I mean I didn't like the feeling of ignorance. Here I was, a Columbia B.A. with a Harvard Ph.D., and supposedly a good chemist. And I didn't know what there was in apples! And even when I heard what was there—Alar, daminozide—I didn't know what these were. I was not happy with myself for not knowing, I was not happy with the apple producers for putting those chemicals in and not letting me know about it. I was not happy with my education for withholding this information.

To take the view that even if *we* do not know, someone *else* knows, and that we should trust that someone else to ensure our health is *naive, unscientific,* and *undemocratic.* Undemocratic because it is not only our right to know but, more importantly as citizens (especially as citizens to whom society has given a free graduate education in chemistry), it is our *duty* to know. If the chemist doesn't know, who then will?

The judgment of naïveté is based on history and knowledge of human nature. The great majority of producers and merchants are scrupulous as far as safety of their products goes. Their reputation depends on their care. But there are also ample examples to the contrary, from stories in the Bible to the Beech-Nut baby food scandal and all those spills in the shipping channels around New York.

To believe that someone *else* knows is unscientific, in view of what we as scientists learn very early on—analyze, check, don't trust the label.

42. Science and Technology in Classical Democracy

I want to return here to Classical Athens in its democracy and reflect a little on how it might have dealt with the Alar issue. There is no question that the matter of a potential public danger, justifiable or not, would have been a subject that the *ekklesia* (the general assembly of citizens) would discuss. The participatory political process ensured this. Pericles' funeral oration, as reported by Thucydides, summarizes the essence of the process and moves us to the connection with science. He says:

> Our ordinary citizens, though occupied with the pursuits of industry, are still fair judges of public matters; for, we alone regard the man who takes no part in public affairs not as one who minds his own business but as good for nothing. We Athenians are able to judge all events, and instead of looking on discussion as a stumbling block in the way of action, we think it an indispensable preliminary to any wise action at all.[1]

It is clear that the citizens of the city-states of Greece felt themselves able to judge, no matter how technical the matter. They gave a place to expertise, to be sure. So military officers, such as the *strategos*, were elected ones, who could hold the office repeatedly. And many, as Pericles, did.

It is interesting to look in the ancient records for attitudes toward science and wisdom. Given the strong technological basis of the success of the Athenian state, its military strategy, weapons, the fast *triremes*, the silver mines, one might have expected more than at least I was able to find. In Aristotle's *Constitution of Athens* one does see mention of contracts to work the mines, leases for the same. These are under the *poletai*, picked by lot from the tribes. There are inspectors of weights and measures, also picked by lot. We have lists of the mining leases and know the appalling working conditions of the miners. We even have a record of an intriguing, in the modern context, proposal by Xenophon to nationalize the privately owned labor force of slaves that worked the Laurion mines.[2]

Athenian silver was won in two ways. Sometimes it came from alluvial white gold, an alloy of gold, silver, and other metals. More commonly, as at Laurion, it occurred in deposits with the lead sulfide, galena. The ore was sorted and concentrated using an ingenious hydraulic system, roasted, and the oxide reduced with charcoal. The crude silver, or lead-and-silver alloy, was then subjected to cupellation. This is an ancient process in which an ore is heated with lead in a vessel shaped of bone-ash and earth. A blast of air oxidizes the nonprecious metals; the base

42.1 An Athenian silver tetradrachm, c. 440 B.C. Reproduced from the catalogue of Sotheby's sale of Greek and Roman coins. Zurich, October 27 and 28, 1993 (Zurich: Sotheby's, 1993), plate 7. Photo courtesy of Sotheby's.

metal oxides dissolve in the lead oxide, which floats on top and is skimmed off. The precious metals, here silver, remain behind.

The matter of ships, the *triremes* that gave Athens its naval power, was a subject of direct interest to the people. The *boule* (the appointed senatelike body) builds new ships, but the people in the *ekklesia* vote to construct them. The people elect the naval architects for the ships, so this is a position of great importance, not selected by lot. I do not know if the naval architects could be reelected, as were the *strategoi*.

But there is little else. Perhaps this is because the records have not survived, or it is conceivable that my search for public attitudes toward science in the old Athens is in vain because much of education, industry, agriculture and trade, and therefore technology, was not a political matter but was left to private enterprise, not debated in public assembly areas.

There is that indelible stain on democracy, the trial of Socrates. Though the final verdict was in part provoked by the philosopher's almost arrogant intransigence, the prosecution itself burns on our conscience. Here is a seeker after wisdom, a questioner if not a scientist, a prophet, silenced by the *people*. Not by one tyrant, but by 280 of his fellow citizens. No wonder that his followers, Plato and Aristotle, looked unkindly on democracy and favored a government of philosopher-kings, of experts. Often scientists join in their dream. But that is just that, only a dream, for reasons I will now explore with you.[3]

43. Anti-Plato; or, Why Scientists (or Engineers) Shouldn't Run the World

Listening in on the easy private banter of scientists, one hears rumors of the new, who's moving where, recitations of funding woes, and, on another level, claims for the rationality of science, the usual deprecation of politicians, and sometimes a disparagement of the seemingly "soft" issues of the arts and humanities. If only the rational approach of science were applied to the way countries are governed, then, ah then, the problems of this world would vanish—or so the argument goes.

Some of this can be dismissed as self-serving fraternal (until recently) camaraderie. But not all—much of it reveals a primitive, flawed worldview, a fallacy that cuts across cultures and political systems. While it is not certain that Plato would allow plebeian scientists as philosopher-kings, some of Plato's naive faith in the supposedly rational surfaces in this contemporary guise.

Modern science is an incredibly successful Western European social invention, an efficient enterprise for gaining reliable knowledge of some aspects of this world, and for using that knowledge to transform the world. At its heart is careful observation, of nature and of our inter-

This chapter is adapted from Roald Hoffmann, "Why Scientists (or Engineers) Shouldn't Run the World," *Issues in Science and Technology* 7, no. 2 (1991): 38–39.

ventions in it. One might be searching for the molecule that gives Tyrian purple its color, or how one could modify that molecule to achieve a more brilliant purple, or a blue.

The world of the scientist is one in which complexity is simplified by decomposition. This, as much as mathematization, is what I mean by analysis (of the nonchemical kind). Discovering or creating (in chapter 9 I've argued for restoring the primacy of the latter metaphor), the scientist typically defines for him or herself a universe of study in which the outcome may be intricate and surprising, but in which there is no doubt that an analysis is possible. There *is* a solution: the dye in royal purple *has* a structure; there must be *reasons* for the limited ability of pandas to breed in captivity. Scientists admit that there may be several factors contributing to one observable or effect; but no matter how complicated, these can be analyzed and taken apart by clever, appropriately trained, initiates communicating in the universal language, broken English.

Contrast this carefully constructed world of the scientist at work with the haphazard reality of emotions or human institutions. Is there a single cause for that young man's crack addiction? Why did brothers kill brothers in the American Civil War? Or do so in the former Yugoslavia? What is the logic of romantic love? Should we have affirmative action programs? Much of the world out there is intractable to simplistic (or even complex) scientific analysis. That world, life itself, *is* subject to ethical and moral debate, to claims of justice and compassion. A clear statement of issues, alternatives, and consequences helps, as may the sometimes aimless dialogue in which contending ethical stances are voiced, and people get off their chests what they must. This is the catharsis that makes participatory democracy work. The resolution of personal and societal problems is not achieved by scientistic claims that a unique rational solution exists.

Scientists, in my experience, are prone to such claims of speaking for rationality. They see that careful analysis works in their research. Confused, even hurt, by the complexity of the world we live in, we reach, naively, for the dream that the wild universe of emotions and collective actions is governed by some rational principles, still to be discovered. Curiously, religion, which science was supposed to supplant, offers a similar (and to me, personally unsatisfactory) worldview. We tend to see the world in black and white, wishing that the gray areas which push into our consciousness in every moment of real life would just go away. If only the doers and makers of the real world (the worst of whom we'll call politicians) would listen to us, then the world would run right.

Well, we have recently witnessed the failure of one such scientistic or technocrat-run dream—Marxism. Whatever culture it has overlain—whether Russian, Chinese, or Cuban—Marxism has proven itself economically unworkable and has perverted its underlying just social core by showing itself to be infinitely corruptible. Scientists don't like to hear this, but Marxism was a "scientific" social system. Marx and Engels drew upon a tradition that forecast a science of society. Their socialism was powered by the myth of infinite progress, cast in the capability of man to transform society as he had transformed nature.

So . . . if not running the world, where should scientists be? It seems to me that scientists are at their best when they are out of power but still engaged in the political process. Then they are motivated to speak as the voice of reason, to give sound advice, to counter ascendant irrationality. Their competence meshes with the demands of the role they play. But were they in command, I think the *hubris* that they, and only they, are reasonable, is likely to lead them to unfeeling excess.

I exaggerate, I know. If scientists are to be faulted, it is for their insufficient participation in the political process. Once they enter the arena, they are no better than others engaged in politics. And no worse. There is, for example, a tradition of scientists and engineers in French politics, from Lazare Carnot and his grandson Sadi Carnot to a postdoctoral student of mine, Alain Devaquet. And neither Margaret Thatcher's shortcomings nor her achievements are to be credited to her undergraduate degree in chemistry.

44. A Response to Worries About the Environment

An editorial by Philip H. Abelson in *Science* summarizes one attitude toward our society's concerns about the environment. Entitled "Toxic Terror: Phantom Risks," it begins and ends as follows:

> The public has long been subjected to a one-sided portrayal of risks of environmental hazards, particularly industrial chemicals. Only a few individuals have attempted to bring balance into the picture. They have faced a self-serving, formidable de facto alliance of media, well-heeled environmental organizations, federal regulators, and the plaintiff's bar. . . . Reviews of the accumulating history of the utterances of the doomsayers reveal their lack of judgment, respect for facts, and honesty. Their assertions are not a sound basis for wasting trillions of dollars on phantom risks.[1]

Whereas I have valued highly the democratizing, progressive nature of chemistry, statements of the type quoted strike me as being *so* far off the mark. They miss the psychological and moral center of all of our concerns about the environment, and furthermore they show an unhealthy attitude toward the democratic process.

It isn't easy to find a middle ground in this confrontation. Let me try for one. What is, or should be, the proper response of chemists to environmental concerns? I believe that response must involve:

1. The recognition that these concerns are based both on technical risk *assessment* (the "facts") *and* on risk perception (psy-

chological, often subjective). And that these ways of evaluating risk (which I will try to distinguish) may not coincide.

2. A realization that in devising the controls that a democratic society imposes on unavoidable risks to person and property, the psychological *perception* of risks figures legitimately, whether we like it or not.

3. The fact that democracy demands a platform for countervailing opinions, and that environmentalist attitudes are clearly within the range of what is acceptable.

The *assessment* of risks is not easy. It involves centrally analytical chemistry and chemical instrumentation. It requires great ingenuity, which we have as a profession given, in the design of schemes, scales, and chemistry to reliably detect substances at unimaginably small levels.

Risk *perception,* as I see it, is not just technological risk assessment, a matter of spelling out the hazards as best as we know. There is a strong psychological component to risk perception, and empowerment figures prominently. By empowerment I mean the reality *and* perception that the person undergoing the risk has some control over the risk.[2]

I suspect empowerment plays the dominant role in personal judgments of risk. We feel safer driving a car rather than flying in an airplane, despite accident statistics to the contrary. We keep on feeling safe, most of us, even if we have drunk a little alcohol. Why? Because it is we who are driving, but someone else is flying the plane. Much of the fear of nuclear power generation and of other technological dangers, real or unreal, derives not so much from ignorance of the processes as from the feeling that we are not in control of the situation.

Empowerment requires access to knowledge *and* a democratic system of government. Even the best of present systems of governance are just an approximation to the ideal of democracy. Still, no amount of knowledge, no matter how skillfully and widely taught, will assuage fear of the synthetic unless people feel that *they* have something to say, politically, about the use of the materials that frighten them.

What I say here is not radical but the common opinion of experts on risk. Here is what Peter M. Sandman, director of the Environmental Communication Research Program at Rutgers, says:

> When you have a public that is both informed and empowered, it is more reasonable. . . . It's not that an informed public tolerates more risk; it chooses better which risks to tolerate. But an informed public without being empowered or explanations without a dialogue have next to no value.[3]

Sandman points to "outrage factors," all the psychological components of risk perception. Let me choose some from among many he enumerates:

- *Voluntariness:* A voluntary risk is much more acceptable to people than a coerced risk, because it generates no outrage. Consider the difference between getting pushed down a mountain on slippery sticks and deciding to go skiing. . . .
- *Morality:* American society has decided over the last two decades that pollution isn't just harmful—it's evil. But talking about cost-risk tradeoffs sounds very callous when the risk is morally relevant. Imagine a police chief insisting that an occasional child-molester is an "acceptable risk." . . .
- *Diffusion in time and space:* Hazard A kills 50 anonymous people a year across the country. Hazard B has one chance in 10 of wiping out its neighborhood of 5,000 people sometime in the next decade. Risk assessment tells us the two have the same expected annual mortality: 50. "Outrage assessment" tells us A is probably acceptable and B is certainly not.[4]

Is there anything wrong in the enacting of legal codes based not only on technical risk assessment but also on a moral *perception* of the risk? I don't think so—the law has always had a consensual moral as well as a material basis. If you don't like that, I ask you to conceive of arguing before a congressional committee on that acceptable rate of child molestation or in favor of euthanasia of the physically impaired elderly.

Before I leave the subject of empowerment, I want to salute the ancient Greeks not only for their philosophy but also for their ability to devise ingenious social structures that truly gave citizens the sense of being empowered. The large juries, the *boule*, the *dikasteria*, the rapid rotation of offices by lot—all of these kept everyone involved. Some of the Athenian inventions that have lapsed need to be revived, for instance the *euthuna*, the detailed investigation of an officeholder's conduct at the *end* of his term of office. It was and is a wonderful idea, and it would be good to have it become a routine procedure for anyone who might profit in power or wealth from his or her office.

I want to return to an attitude toward environmentalists. Some chemists think that the environmentalists' fears are irrational. Simple psychology tells us that in addition to reason and empowerment, even before them, compassion figures prominently in responding to and allaying any fears. Friends, chemist friends, if someone comes before you verbalizing anxiety over a chemical in the environment, don't harden your hearts and assume a scientistic, analytical stance. Open your hearts, think of one of your children waking at night from a nightmare of being run over by a locomotive. Would you tell him (or her), "Don't worry, the risk of you being bitten by a dog is greater"?

Not that environmentalists are children. Within two centuries—the centuries of chemistry—science and technology have transformed the world. What we have added, mostly for the best of reasons, is in danger of modifying qualitatively the great cycles of the planet. The amount of nitrogen fixed from the atmosphere by the Haber-Bosch process, that masterpiece of chemical ingenuity, is, I suspect, comparable to global biological nitrogen fixation.[5] These changes have been wrought in the geological equivalent of the blink of any eye. Gaia may have the restoring forces to deal with our transformations, but the world that results could be one in which humankind will not play a role.

44.1 Shell oil refinery at dawn. In background is Mt. Baker, Anacortes, Washington.
(Photo by Richard During/TSI)

We see the effects of our intervention in the change in the ozone layer, in the pollution and acidity of our waters, in why we wash an apple, in the crumbling statuary—our heritage, dissolving in the open air. There is a good reason why the original of Michelangelo's *David* was moved off the Piazza della Signoria in Florence. There are very good reasons why we should wake the environmentalist within all of us.

45. Chemistry, Education, and Democracy

To me the Alar controversy was humbling, educational, and instructive, an opportunity to learn rather than a chance to blow off some steam against environmentalists. I learned some chemistry from it; I learned some from Bhopal, and I intend to learn some from the next chemical disaster. People's minds open up when knowledge is accompanied by a relationship to something critical—a disaster, one's body, even the prurient and scandalous. One can use ill events in an educational sense.

I have come to education. I view education as a crucial part of the democratic process, a privilege and a duty of the citizen. In fact, I'm not concerned about scientific illiteracy (and this is my opinion only, I remind you) so much from the point of view of its limiting our man- or woman-power base or affecting our global economic competitiveness. What worries me about prevalent chemical illiteracy—a failure of the educational process—are two other matters.

First, if we do not know the basic workings of the world around us, especially those components that human beings themselves have added to the world, then we become alienated. Alienation, due to lack of knowledge, is impoverishing. It makes us feel impotent, unable to act. Not understanding the world, we may invent mysteries or new gods, much as people did around lightning and eclipses, around St. Elmo's fire and volcanic sulfur emissions a long time ago.

My second point of concern about chemical illiteracy returns me to democracy. Ignorance of chemistry poses a barrier to the democratic process. I believe deeply, as must be clear by now, that "ordinary people" must be empowered to make decisions—on genetic engineering or on waste disposal sites, on dangerous and safe factories or on which addictive drugs should or should not be controlled. Citizens can call on experts to explain the advantages and disadvantages, the options, the benefits and risks. But experts do not have the mandate; the people and their representatives do. The people have also a responsibility—they need to learn enough chemistry to be able to resist the seductive words of, yes, chemical experts who can be assembled to support any nefarious activity you please.

Here then is the importance of constructing primary and secondary school chemistry courses that reach out to a wide audience. And of training and rewarding teachers that can teach these classes. Chemistry courses must be faithful to the intellectual core of the subject. But they also need to be attractive, stimulating, intriguing. They must be aimed primarily at the nonscience student, at the informed citizen, not toward the professional. New chemists, brilliant transformers of matter, will come from among these youngsters. Of this I'm sure. But they will not be able to do what they are capable of unless we teach their friends and neighbors, the 99.9 percent who are *not* chemists, what it is that chemists do.[1]

PART NINE

The Adventures of a Diatomic

46. C₂ IN ALL ITS GUISES

I love this molecular science. I love its complex richness, and the underlying simplicity, but most of all the life-giving variety and connection of all of chemistry. Let me give you an example of what I like, because we have wandered far from the beauty of the beast. While I have chosen to look at chemistry in terms of themes such as analysis, synthesis, and mechanism, the classical subdisciplines of organic, inorganic, biological, physical, and analytical chemistry have a persistent life of their own.[1] I like the chemistry that subverts these divisions.

C_2 is a simple diatomic molecule. Just two carbon atoms. It's not very stable, quite unlike the familiar O_2, N_2, F_2. But whenever an arc is struck between two carbon atoms, one gets a little C_2 (and a little soccer-ball-shaped C_{60}, a so-called buckminsterfullerene; but that's another, quite marvelous, story)—enough to obtain its structure by one of those spectroscopies I mentioned earlier. There is also a good bit of C_2 in comets. And C_2 is responsible for the blue light we see in flames.

You will ask, "What structure does the C_2 molecule have?" The molecule looks like a dumbbell, and the only free variable it has is the distance between the two carbons. That distance is 1.2425 Å (1 Å or Ångström is 0.00000001 cm, 1 to 3 Å is a characteristic distance between bonded atoms in a molecule) in the ground state, the stable form of the molecule.

Any molecule, C_2 as well, also exists in so-called excited states. These result from absorption of light by the molecule, or from the input of energy in other ways. Molecules do not persist forever in their excited states, but after a certain amount of time (which may be minutes or microseconds) they return to the more stable ground state, sometimes emitting light in the process. There are C_2 molecules produced in normal flames—they derive from the carbon-containing fuel, and they eventually end up as CO_2 or soot. In the flame they are produced in an excited state, their excitation deriving from the heat of the complex reactions going on in the flame. Returning to their ground state, they emit blue light.

One way to describe a diatomic molecule is by a "potential energy curve." This is a graph showing how the energy in the molecule varies as a function of separation between the atoms. Illustration 46.1 shows one such curve; the energy is plotted vertically, the separation between atoms (in Å) horizontally.

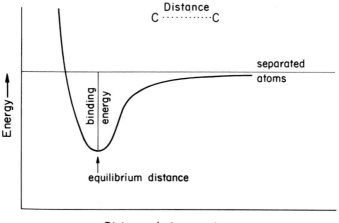

46.1 A "potential energy curve" for a molecule.

Translated into English, the curve says: The energy goes down for a while as the atoms approach each other. Then it rises, eventually with a vengeance.

The distance at which the curve changes from going down to going up (the point of lowest energy) is called the "equilibrium distance" of the molecule. And the energy by which it is lower (more stable) with respect to the separate atoms is called the "binding or dissociation energy." The well in the potential energy surface describes the molecule—the molecule "sits" in that well. C_2 has a distance of 1.2425 Å in its ground state.

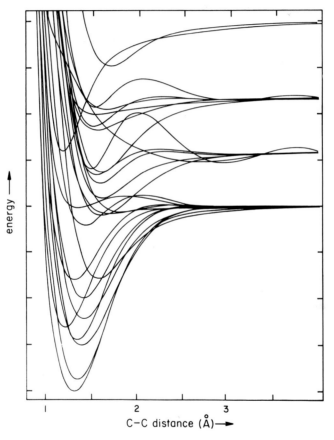

46.2 Computed potential energy curves for C_2.

Each excited state is a thing unto itself, with its equilibrium distance and well depth different from the ground state. Illustration 46.2 shows the potential energies not only for the ground state of C_2 but also for many of its excited states.[2]

Of the multitude of these states that are calculated by theory, no less than thirteen (the ground plus twelve excited states) have been observed experimentally. Their equilibrium C2DC distances are indicated below (the part Greek name labels are descriptors carrying some information about the state):[3]

Table 4
C_2's Thirteen Experimentally Observed States

State of C_2	C-C Distances (A)
$^1\Sigma_g^+$	1.2425 (ground state)
$^3\Pi_u$, $^1\Pi_u$	1,3119, 1.3184
$^3\Sigma_g^-$	1.3693
$^3\Pi_g$, $^1\Pi_g$	1.2661, 1.2552
$^3\Sigma_u^+$, $^1\Sigma_u^+$	1.23, 1.2380
$^3\Pi_g$	1.5351
$^1\Sigma_g^+$	1.2529
$^3\Sigma_g^-$	1.393
$^3\Delta_g$	1.3579
$^1\Pi_u$	1.307

Note the range of C-C distances—from 1.23 to 1.53 Å. A chemist reader might also notice a remarkable fact—this molecule has an excited state which has a shorter C-C bond than the ground state. That's extremely rare, but it has its explanation in the electronic motions, the so-called molecular orbitals which describe the quantum mechanical states of the electrons in the molecule. (This happens to be what I make a good living at calculating badly.)

The study of the excited states of C_2 falls clearly in the realm of physical chemistry. Now let's switch to a group of three molecules representative of much of organic chemistry (and, incidentally, all of commercial importance). These are ethane (C_2H_6), ethylene (C_2H_4), and acetylene (C_2H_2), shown in illustration 46.3. Ethylene is produced in a staggering amount (41 billion pounds in the United States alone in 1993). These molecules are the archetypical C-C single, double, and triple bond. And, as one might expect, the stronger the bond, the shorter it is. Note the range of C-C bond distances in these molecules, which is just about the complete range of distances in the millions of organic molecules we have wrought—it is between 1.21 and 1.54 Å.

That's not very different from the repertoire of distances the ground and excited states of C_2 display. Could that be an accident?

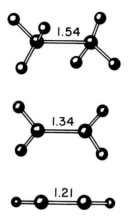

46.3 The archetypical organic molecules: from top to bottom—ethane, ethylene, acetylene. C-C distances in Å = 10^{-8} cm are indicated.

We move on to organometallic chemistry, a lovely interface between organic and inorganic chemistry that has exploded with activity in the last thirty years. Illustration 46.4 (left) shows one organometallic molecule, made by my Cornell colleague Peter Wolczanski and his coworkers. It has a C_2 unit neatly bridging two tantalums, each of which has some bulky molecular bushery (not all shown here) around it.[4] Illustration 46.4 (right) shows another organometallic molecule, one made by Michael Bruce's group in Adelaide, Australia. It has four rutheniums caught in huddling around a C_2.[5]

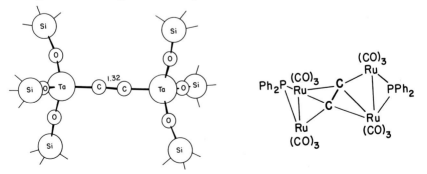

46.4 The compound at left is $[(t-Bu_3SiO)_3Ta]_2C_2$, t-Bu=$C(CH_3)_3$; the one at right is $Ru_4(C_2)(PPh_2)_2(CO)_{12}$, Ph=$C_6H_5$.

We have crossed the bridge to inorganic chemistry, a distinction that seems to matter to some. A group at Milan, Italy, has been prolific at making metal clusters. In illustration 46.5 you see such a cluster: seven cobalts, three nickels, many surrounding carbon monoxides—and smack in the middle of the cage is a C_2, with a middling bond length of 1.34 Å.[6]

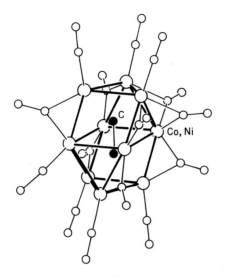

46.5   The $Co_3Ni_7C_2(CO)_{15}^{3-}$ cluster.

Once you've lit a carbide lamp, you never forget the smell of wet acetylene. Illustration 46.6 shows the structure of calcium carbide, CaC_2. Union Carbide began by making this molecule. On adding water it yields acetylene, which burns in a carbide lamp.

Calcium carbide is called an extended structure, a crystalline solid. It is made of atomic or molecular units, marching off regularly to infinity (or partway there). The C_2 units that you see so clearly in this structure have a very short bond length of 1.19 Å.

We have now crossed from inorganic to solid state chemistry. The solid state encompasses a wide variety of chemical compounds, mostly inorganic. It includes minerals, catalysts, high-temperature superconductors, metals, magnets, alloys, glasses, ceramics, and much more.

Here is another typical solid state structure, made by Arndt Simon's

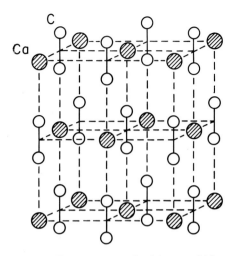

46.6 The structure of calcium carbide.

collective in Stuttgart. $Gd_{10}C_6Cl_{17}$ is the kind of structure we wouldn't dare to show beginning students in chemistry; we like to shield beginners from this beautiful complexity (illustration 46.7).[7] The molecule contains no less than seven octahedra of gadolinium, surrounded by assorted chlorides. Within each and every octahedron resides a C_2 unit!

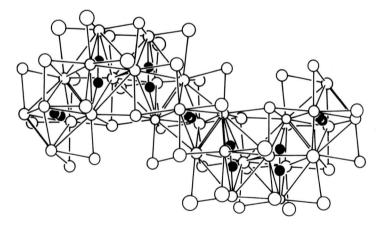

46.7 The $Gd_{10}C_6Cl_{17}$ structure.

Let's look at one more structure. When organic molecules are put down on clean metal surfaces, they are often ripped to pieces. This is not as bad as it sounds, for they are then reassembled to form other molecules; metal surfaces often act in just this way as commercially important catalysts. On a certain silver surface, Robert Madix and his co-workers at Stanford have found that acetylene, C_2H_2, decomposes to precisely our friend the C_2 unit, which then sits on the surface as shown in illustration 46.8.[8]

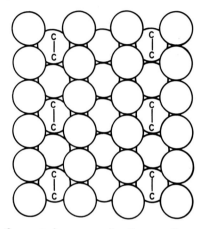

46.8 Suggested structure for C_2 on a silver surface.

In illustration 46.9, I have redrawn all of these molecules in a hub, centered by C_2. Now these structures come from different parts of the molecular enterprise: physical, theoretical, organic, organometallic, inorganic, solid state, surface chemistry.

I think what nature is telling us, as clearly as possible through this staggering richness, is, "You guys (including women, who now make up 23 percent of the Ph.D.'s in chemistry in the United States)[9] may divide chemistry as you wish, but I'm telling you that the world is one. There are C_2 molecular units in each of these structures, acting out a dance of varying distances."

I think this is beautiful.

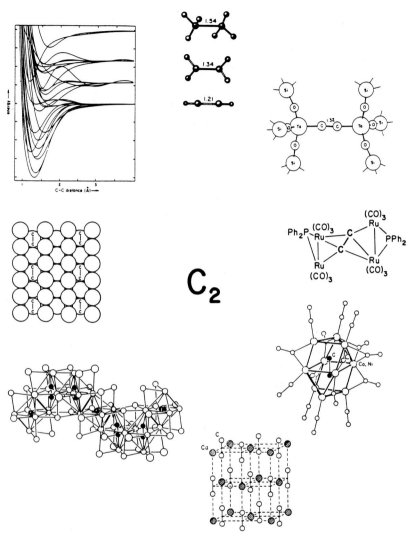

46.9 The C$_2$ wheel.

Part Ten

The Dualities That Enliven

47. CREATION IS HARD WORK

What are the tensions we've seen in chemistry? First and foremost, there is the question of identity, of *being the same and not the same*. We've learned to distinguish between molecules that differ only in such exquisitely fine detail—for instance, in being mirror images of each other. With all the intricate complexity of good theater, chemists have written a few new scenes for an old strategy that evolution first came up with, molecular mimicry. And they have designed some spectacular life-saving drugs. What are you? Who are you?

Synthesis is the partner of *analysis* in chemistry. I would claim, in fact, that synthesis deserves primacy (if one were inclined to assign such) as the single most chemical of chemical activities. Analysis (what Döbereiner taught Goethe) is common to other sciences. It is the scientific method of work. But chemistry is unique, for central to it is the making of things, just as much as taking them apart. Too little attention has been paid by reductionist philosophers to how this very chemical procedure, synthesis, works.

Creation and discovery are nicely balanced in chemistry. And closely related to the synthesis and analysis theme. Chemists create (subject to rules that we are still discovering) new molecules. We create, not only new molecules but new ways—in a way new rules—for making them.

The intriguing thing about creation is that it is so intellectual and so down-to-earth practical. Creation is such hard work.

Because the molecule is central to chemistry and because the molecule is, on the average, an aggregate of atoms of fixed geometry, *structure and its representation* is another of our continua. We could identify the associated struggle as that of the *ideal* versus the *real*—or else focus, as I have done, on the range of representations of chemical signs. These are part iconic (looking like the thing they represent), part symbolic (where the only connection is by convention, by mutual agreement, on a sign devoid of pictorial content).

That molecular structure has to be represented leads quite naturally to the next subject I discussed, which is the nature of the chemical article. This seemingly dull, ossified, ritual method of communication is full of tension, between what is to be *revealed* and what *concealed,* between the mode of expression (*dispassionate*) and the intent (*impassioned* rhetoric). Chemists cope with the article, thus communicating reliable information and building reputations. Even style is possible (pretty incredible given the limited literary palette allowed).

Returning to synthesis, the archetypical chemical action, it is easy to see how the *natural/unnatural* dichotomy rears its head. We make in the laboratory immensely complicated molecules that occur in biological organisms. And we also synthesize molecules (such as cubane) that are very simple in appearance but devilishly hard to make. Chemistry bridges the natural/unnatural gap, or, better said, subverts it at every turn. Still our psychology struggles, for an assortment of reasons, some "good," some "bad," to maintain a separation of the natural and artifactual. Chemists need to be sensitive to that.

After analysis and synthesis, the next most typical chemical activity (which extends human curiosity and merges with what motivates history and psychology) is the *study of mechanism.* How did (does) that reaction happen? When it works, the elucidation of the mechanism of a chemical reaction is the discovery mode of science at its best—except that it doesn't always work according to the textbook model, and psychological factors intrude.

Static and dynamic is another tension underlying the molecular science. The air may seem quiescent, but in reality it is a mad three-dimensional dance floor, in which solitary molecules cruise at the speed of sound, but not very far, before they collide with another. From those collisions comes chemical reactivity, and another seemingly static situation, equilibrium. This absolutely natural balance is based on the frantic motion of reactants to products, and products to reactants. Chemical equilibrium also is characterized by a macroscopic, almost lifelike resistance to our selfish perturbations.

In the life of a single great chemist, Fritz Haber, we can see played out many of the creative tensions of chemistry. Haber moved between industrial and academic chemistry. He discovered a process for synthesis (of ammonia) on an unparalleled scale, using his knowledge of the mechanism of the reaction. He failed, at another curiously alchemical point in his life, because of a faulty analysis. He also failed, morally this time, at still another juncture by putting his creative chemical abilities at the service of a cruel military (but militarily ineffective) innovation—poison gas. And the world around him changed, so his identity was vetted—Haber was no longer the same, a good German. In 1933 he became what he felt he was not, a Jew.

Utility and harm—to individual chemists, to our fellow citizens, to the world that we will leave behind for others—is another axis along which any chemical activity must be examined. I did this for Haber, making judgments with which not everyone would agree. Concerns about the environment make it imperative that chemists see the world not in the black and white of supposed reason but also with compassion for the moral and psychological concerns that bother us all. Curiously, people are often interested in chemistry for reasons that derive from the same psychological factors that make them fear, seemingly irrationally, technological catastrophe. And any given molecule (e.g., ozone, nitric oxide, or morphine) may be both Dr. Jekyll and Mr. Hyde.

The final dichotomy is embedded in the nature of scientists themselves, and not just chemists. That is that we are fated to act, which means to create. The consequences of that creation may be good or bad. It is not easy to be a socially responsible scientist. But then it is never easy to be human.

48. Missing

There are several important dualities that I have barely touched on in this book. One is what Thomas Kuhn has called "the essential tension" between paradigmatic work (call it productive routine) and revolution.[1] Anxious as we always are to live out our myths, the popular image of the scientist is as an uncompromising innovator, ever open to new ideas. Kuhn argues cogently for an acceptance, even valuation, of the contrasting reality that most science is, and should be, paradigmatic. He says: "the productive scientist must be a traditionalist who enjoys playing intricate games by preestablished rules in order to be a successful innovator who discovers new rules and new pieces with which to play them."[2]

Organic synthesis is a wonderful example of the working out of Kuhn's essential tension. One tries old reactions; some work, some don't. So, driven by the goal of creating a molecule never made before, one finds new reactions, which very quickly become part of the organic chemist's standard repertoire.

Still another tension that I have left largely unexplored is that of *trust* versus *suspicion*. Remember all those references in scientific papers? Some are window dressing, to be sure. But most are signs of trust, a dependence on what went before, perhaps a partial listing of the giants on whose shoulders we stand.[3] Those references are the basis of an information industry and a "scientometric" tool—citation indexing.[4] And they are the most direct source of the satisfaction that scientists feel in their work. There is no better feeling than to see your work cited amply by people you do *not* know.

References, especially to known facts and methodologies, reflect a great degree of trust in what has been published before. But that trust is always tempered with suspicion. So the chemist who buys CH_3CD_3 will find a way to analyze it before using the labeled ethane in a crucial experiment. The vaunted reproducibility of science takes a real beating in the chemical laboratory. Here is what a leading chemist of our time, Robert G. Bergman, reports:

> My research interest is in the study of reaction mechanisms—finding out how chemical reactions work. In order to do this, we normally need to synthesize particular compounds whose molecules have consciously chosen structural features. It is therefore common for a person in my research group to start a new project by repeating (or attempting to repeat) a literature preparation of an organic or organometallic compound whose synthesis has been published in the literature, and then utilizing that material in a new chemical transformation.
>
> The startling fact is that almost half of the literature synthetic procedures we attempt to repeat initially fail in one way or another—that is, they cannot be carried out, to give the yield of product claimed, by following only the directions described in the published paper. A reasonably large fraction of these "recipes" *can* be reproduced after modification or discussions with the initial author. Some, however, cannot be repeated in our hands no matter what we do.[5]

This lack of reproducibility is certainly not limited to Bergman's group. He goes on to cite the confirming evidence of two journals which publish syntheses that are deliberately checked before publication.

At this sometimes tense border of trust and suspicion we all operate. Incredibly, the system works very well.

I have also not discussed a duality inherent not only to chemistry but to most science, and this is *observation* versus *intervention*. The problematic here has many guises, from the Heisenberg uncertainty principle to the difference between in vivo and in vitro biological studies. On the subatomic level, observation is intervention; the energies involved in the act of observation may perturb what is being observed—this is Heisenberg's uncertainty principle. In chemistry intervention and observation are intricately linked; observation (for instance, a new reaction serendipitously discovered) is almost immediately followed by intervention (an attempt to perfect that reaction by changing conditions, or to modify it).[6]

Another duality is *pure/impure*, critical to the question of identity of chemical substances. As previously noted, nothing is pure—and there are good reasons for that, having to do with entropy and evolution. The struggle to define the approximate identity of two nearly pure mixtures is fascinating (as is the apparent confrontation between chemistry's acceptance of the impure as natural and religion's clearly voiced aspiration for the moral good of purity).[7]

There are still other dualities we might have explored. Gerald Holton has written about "themata" of science. These are intellectual categories or ways of thinking about any particular piece of science. They could be thought of as axes. Holton demonstrates that such themata recur in the work of various scientists, that a certain outlook on science, as defined by the point on the axis where one chooses to be, is set early, and is firmly held to by many scientists.[8] And others. Among Holton's themata are:

analysis—synthesis
constancy—change
many—one
complexity—simplicity
parts—whole
mathematics—materialistic models
disaggregation—aggregation
representation—reality
reductionism—holism
discontinuity—continuum
dismemberment—unification
differentiation—integration

It is evident that some of these dualities are just the same as I have discussed. Others would serve as a useful point of departure, as good as some I've chosen.[9]

Henning Hopf, a perceptive observer of the human and chemical sciences, has also remarked that certain oppositions have a venerable history in chemistry. There is no contrast as seminal to chemists (and as difficult to quantify in its manifold manifestations) as that of acid and base. Polarities of attraction versus repulsion, softness versus hardness (of acids and bases), electrophilicity versus nucleophilicity, covalent versus ionic bonding—all have defined discourse in chemistry in this century. These are technical concepts, to be sure. But they point to differences that have fascinated chemists.

49. An Attribute of the Devil

is what Wolfgang Pauli called dichotomizing.[1] It is too easy—and ultimately tiresome. But there is a difference between just listing oppositional qualities (good/evil, symmetry/asymmetry) and coming to terms with the tense synthesis of opposites that makes life interesting. So neither men nor women are exclusively good or evil; and if beauty is to be found, it surely is at the edge where symmetry and asymmetry contend.

There is a philosophical perspective that seems (on the surface, at least) close to the road I've followed in analyzing chemistry—the Hegelian dialectic. Hegel proposed a mode of argumentation that was also a prescription for understanding. For any thesis there is an antithesis. From the contention of the two there evolves a synthesis (we can't get away from this word!).[2]

A way of polarities or dualities certainly has a dynamic that resembles the working of Hegel's dialectic. But I think my approach of looking at chemistry moves beyond dualism, in two ways. First, the chemical fact, or the act of the chemist on obtaining that fact, is a balancing act on the polar axis—a different compromise for every molecule and for every maker of that molecule. Second, there is no real single thesis or antithesis but rather a multiple perspective, if not cubist at least multidimensional.[3] A molecule may be similar to another one, harmful or beneficial, discovered or created, quiescent or in rapid motion. But, under certain conditions, it may also be all of these!

Why opposition? I think there is little choice but to focus on polarities in describing a living, changing *human* activity such as chemistry.

Here is what Emily Grosholz, a poet and philosopher, says in her remarkable essay on nature and culture in two books by W. E. B. Du Bois:

> Metaphysical schemes that accord reality to development must display the structure of reality in terms of possible changes; change requires difference, and difference takes the form of binary oppositions in our language and thought. The venerable binary oppositions of metaphysics are a part of human wisdom; for all their indeterminacy, they stand for something fundamental and inescapable.[4]

The oppositions I've chosen reflect the *life* of chemistry. And they gain strength through the subconscious associations we make of science and individual psychology.

It's no accident, I think, that there is a special pull on us, an archetypal feeling, exercised by that classic of duality, Robert Louis Stevenson's "The Strange Case of Dr. Jekyll and Mr. Hyde" (1886). Hiding in that story of identity is also a critical chemical duality:

> My provision of the salt, which had never been renewed since the date of the first experiment, began to run low. I sent out for a fresh supply, and mixed the draught; the ebullition followed, and the first change of colour, not the second; I drank it, and it was without efficiency. You will learn from Poole how I have had London ransacked; it was in vain; and I am now persuaded that my first supply was impure, and that it was that unknown impurity which lent efficacy to the draught.[5]

49.1 "And as I looked there came, I thought, a change . . ." An illustration by William Hole to Robert Louis Stevenson's "The Strange Case of Dr. Jekyll and Mr. Hyde." (Archive Photos)

Avner Treinin, one of Israel's leading poets, is also a distinguished physical chemist. In an essay entitled "In Praise of Dualities," he writes:

> But probably the most compelling source of my attraction to science and poetry has been not the similarities between them but rather the dissimilarities, even contradictions: to see the same thing from two apparently opposite perspectives and feel the mounting tension between the two.
>
> There is something strange in our attitude toward contradictions. From childhood we are told to avoid them, to be consistent, while the whole of our experience teaches us not only that we are concentrated solutions of contradictions, but also that without them nothing could exist. Essentially, this is what dialectics is all about. The atom itself, the building block of all matter, is composed of positive and negative charges, and everything that flows (water, electricity, the pulses in my brain which are now composing this sentence), flows between opposite poles, that is, through a potential gradient. Moreover, from modern physics we learn that the only way to understand reality (if this can be called understanding) is to use two contradictory pictures which complement each other: particles and waves, or mass and energy. . . .
>
> So what is the wonder in finding that poetic and scientific pictures may complement each other in conveying to us some sense and essence of our existence, and that by bringing the two pictures together a powerful spark can be generated in the mind?
>
> As any physical chemist dealing with surface phenomena knows, the important things happen at the boundaries between things, where something ends and something begins, like the tension created between adjacent poles, between body and soul, content and form, particles and waves, number and feeling. It is at the interface between two different media that light is reflected, refracted, converges, stimulates the optic nerve, forms the image—and we see.

In his notebooks, Leonardo da Vinci teaches his students how to paint the Deluge. After referring to many of its horrors, such as ships broken to pieces, flocks of sheep beaten on rock, hailstones, thunderbolts, whirlwinds, rotting corpses, and so on, he adds: "And if the heavy masses of ruin of large mountains or of other grand buildings fall into the vast pools of water, a great quantity will be flung into the air and its movement will be in a contrary direction to that of the object which struck the water, that is to say: the angle of reflection will be equal to the angle of incidence." Here is the confrontation between the "cold" physical law of reflection (the equality of angles) and the highly emotional description of death and destruction, between the concrete and the abstract, the general and the particular, the reproducible and the irreproducible, between order and chaos—between science and poetry. It is a very intense confrontation which strongly moves the soul. If dualities did not exist, we should have invented them, provided we could do it without any duality to begin with. This is probably the reason why God split Adam into two opposite poles. He wanted him to move, to be alive.[6]

And in a quite different context, the anthropologist Kathryn S. March concludes a paper on "Weaving, Writing, and Gender," in which she discusses the way weaving and Buddhist writing shape and are shaped by gender in the Tamang (a Tibetan origin group in north central Nepal), in the following way:

> Gender as a symbolic system specifically represents this very problematic or paradox, indeed, antinomy: to represent things that are, and are not, the same; things that might be the same if they were not interpreted from opposing perspectives; perspectives that emerge as opposed because they arise as women and men consider the gender logic of each other's position; men and women who, as they consider one another, confront the many ways in which they are, and are not, the same.[7]

50. Chemistry Tense, Full of Life?

So what is chemistry? Is it just the science that we notice only when a tank car of benzene finds itself in the river and a town must be evacuated? Whose most exciting manifestation might be the fireworks on the Fourth of July? Or could this science indeed be lively *and* intellectually deep?

Or is what I've done just a structural device, a trick? Take anything seemingly dull in this world, say, a day in a provincial accounting firm or a hard day spent cutting sugar cane in Cuba. Survey the limits which shape the middle, polarize, dichotomize, deconstruct all peaceful existence as a precarious struggle. If you are convincing enough, you can create tension where none was there before.

I don't think I've stirred up a Potemkin storm. Before there was science, the miracle of substances changing (today we would say "molecules reacting") had a most powerful hold on the human imagination. I refer to alchemy, a cross-cultural activity in which a philosophy of change was conjoined with protochemistry (with, admittedly, some charlatanry thrown in). Chemists would like to forget the esoteric philosophy, keep the protochemistry, and laugh at the charlatanry. But these aspects were all tied firmly together.

The reason alchemy captured the imaginative faculties of people

50.1 Fireworks, a most chemical art. Red comes from strontium, calcium, and lithium salts; blue from copper salts; white from metallic magnesium and aluminum; gold sparks from iron filings; green from barium salts. (Photo by Sepp Dietrich/Tony Stone Images)

over centuries and across cultures is that it touched something deep. Change (and stability) is physical and psychical; juxtapose any two manifestations of change, and one immediately becomes a metaphor for the other.[1]

50.2 An alchemical illustration, from Basil Valentine, *The Twelve Keys: The Hermetic Museum*, 1678.

Goethe's novel, *Elective Affinities,* has been mentioned several times in this book. With good reason, for it is one of the few successful literary works taking its main theme from a chemical theory. The idea of elective affinities—a theory soon superseded—was that certain chemical entities (we would now say molecular fragments) possess a special, definable, chemical affinity for each other. And yet Goethe knew that he had done more than clothe a chemical theory in beautiful language. In an advertisement in *Cottas Morgenblatt* (a newspaper of the day) he explained that the title of his book was a chemical metaphor whose "Spiritual Origin" his novel would demonstrate.[2]

I think chemistry is interesting to its toiling practitioners, and to people who use it (or abuse it) without being chemists, because its activities parallel deep avenues in our psyche[3]—which I prefer to see not as a branching tree of neurons, shaped by genetics and experience (and chance), but as a completely interconnected multidimensional volume. In which a given fact (a molecule, a line from a poem) has a history, a context, to be sure. But it comes to life only if we think of the molecule (or the poem) as suspended—yes, tensely—in a space that is defined by different themes or oppositions.

In an imperfect metaphor, think of the themes as light of different wavelengths. Or think of them as coordinate axes, not very linear, in a multidimensional space. I turn on the light of identity, of the same-and-not-the-same, and I see cubane as different from other C_8H_8 molecules, many of which have been synthesized. I tune to the radiation of cooperation and competitiveness, and before me flashes the image of the Harvard assistant professor who once inspired me to do an overly simplistic calculation on cubane, a man who had devoted years to making the molecule and failed. Had he succeeded, he would have been promoted. I look at cubane in the multicolored light of utility and social responsibility, and I think about whether it should worry us that some of the work on it is supported by military research agencies, or that a derivative has been found to have antiviral activity, or that the strained molecule could find a use as a solar energy storage material.

The different ways in which any molecule is examined, where it falls not on one but on many polarity scales, make that molecule inherently *interesting.* The questions one asks the molecule touch silently—without our knowing it sometimes—on vital questions we should ask ourselves.

51. CHEIRON

No less than four of the signs of the Zodiac are dualistic: Gemini, Libra, Pisces, and Sagittarius. I choose to see the constellations not as a vestige of dark ages but as a timeless pointer to that irrepressible quality of the human soul that eventually led to science—curiosity, and a search for patterns.

Sagittarius is a Centaur. My favorite of these half-man, half-horse creatures of Greek myth is Cheiron. He was the son of Cronus (Zeus's father) and Philyra, the daughter of Oceanus.[1] Immortal Cheiron was sage and kind. From his cave on Mount Pelion he taught the healing arts to Asclepius, the skills of riding, hunting, and pipe-playing to Achilles. He taught Diomedes, who became Jason of the Argonauts, he taught Aeneas. The teacher in me likes this teacher.

Good deeds (and no myth testifies to the contrary) did not bring Cheiron a happy old age. A bystander to a typical Centaur melee which was not his, he was wounded by a poisoned arrow (I wonder what poison) fired by his friend Heracles. The great Centaur screamed in pain, but he could not die, for he was immortal. Finally, Zeus granted him peace, in the process bringing about a significant conjunction of the wise Centaur who taught gods and men, and the rebellious Titan, Pro-

51.1 *The Centaur Cheiron Instructing Achilles,* by Jean-Baptiste Regnault, 1782.
In the collection of the Louvre, reproduced by permission.

metheus, who brought fire to humanity. Prometheus speaks, in the
words of Aeschylus:

> Hear rather that all mortals suffered,
> Once they were fools. I gave them power to think.
> Through me they won their minds . . .
> Seeing they did not see, nor hearing hear.
> Like dreams they led a random life . . .
> From me they learned the stars that tell the seasons,
> their risings and their settings hard to mark.
> And number, that most excellent device,
> I taught to them, and letters joined in words.
> I gave to them the mother of all arts,
> hard working memory.[2]

For this Prometheus was punished, for teaching us to see. He was chained to a peak in the Caucasus, an eagle "feasting in fury on the blackened liver" of the Titan whose name means "forethought."

Hermes, Zeus's messenger, says to Prometheus,

> Look for no ending to this agony
> Until a god will freely suffer for you
> Will take on him your pain, and in your stead
> Descend to where the sun is turned to darkness,
> The black depths of death.[3]

It was Cheiron who was willing to die for Prometheus. And in one of the greatest losses I feel we have suffered, the account of the subsequent reconciliation of Prometheus and Zeus, in the last part of Aeschylus's great trilogy, is missing.

So the fates of Prometheus and Cheiron intersected. The Centaur's name comes from the Greek word for hand, the same word that is at the root of that most subtle difference that may heal or kill, the nearly same, chirality. I imagine Cheiron stretching his hand out to Prometheus, as he gives him the gift of life.[4]

As inherently good as Cheiron was, I do not want to romanticize the Centaurs, who were by and large a rough and immoral bunch. But it is crystal clear that Centaurs are the incarnation of the same and not the same. Man and beast, not whole human, not wild beast. Stationary and fleet, a tenser, complex, yet integrated being. Capable of harm, seeking for the good. Like chemistry.

NOTES

Identity—the Central Problem

1. Lives of the Twins

1. Rosamond Smith, *Lives of the Twins* (New York: Simon & Schuster, 1987), pp. 102–103. Copyright © 1987 by the Ontario Review, Inc. Reprinted by permission of Simon & Schuster, Inc.

2. *Opportunities in Chemistry* (Washington, D.C.: National Academy Press, 1985), p. 136. This graph was adapted from F. L. Johnson, "Advances in the Management of Malignant Tumors in Children," *Northwest Medicine* 7 (1972): 759–64. Dr. Johnson has kindly supplied me with the updated graph reproduced here. See also section on "Cancer Trends: 1950–1985," *1987 Annual Cancer Statistics Review*, NIM publication no. 88-2789 (Bethesda: National Institutes of Health, 1988), pp. II.193–203; M. E. Nesbit, Jr., "Advances and Management of Solid Tumors in Children," *Cancer* 65 (1990): 696–702.

3. Reproduced by permission from M. D. Joesten, D. O. Johnston, J. T. Netterville, and J. L. Wood, *World of Chemistry* (Philadelphia: Saunders, 1991).

4. From *Panta Rhei*, vol. 1 (Lucerne: Hans Erni-Stiftung, 1981), p. 83.

2. What Are You?

1. A. Rapp, "Wine Aroma Substances from a Gas Chromatographic Analysis," in H. F. Liskens and J. F. Jackson, eds., *Wine Analysis* (Heidelberg: Springer, 1988), pp. 29–66. See also Ron S. Jackson, *Wine Science* (San Diego: Academic Press, 1993), chap. 6.

2. F. Jacob, "Evolution and Tinkering," *Science* 196 (June 10, 1977): 1611.

3. Reproduced by permission from *Mineral Digest* 3 (1972): 71.

4. J. P. Marion, F. Müggler-Chavan, R. Viani, J. Bricout, D. Reymond, and R. H. Egli, "Sur la composition de l'arôme de cacao," *Helvetica Chimica Acta* 50, no. 6 (1967): 1509–16.

5. Adapted with permission from W. D. Jones, M. A. White, and R. G. Bergman, "Chemical Reduction of η^5-Cyclopentadienyldicarbonylrhodium," *Journal of the American Chemical Society* 100 (1978): 6770–72. Copyright © 1978 American Chemical Society.

3. Whirligigs

1. See Natalie Angier, "For Insects the Buzz Is Chemical," *New York Times*, March 29, 1994, p. C1.

2. See J. Meinwald, K. Opheim, and T. Eisner, "Gyrinidal: A Sesquiterpenoid Aldehyde from the Defensive Glands of Gyrinid Beetles," *Proceedings of the National Academy of Sciences (USA)* 69 (1972): 1208, and "Chemical Defense Mechanisms of Arthropods XXXVI: Stereospecific Synthesis of Gyrinidal, a Nor-Sequiterpenoid Aldehyde from Gyrinid Beetles," *Tetrahedron Letters*, no. 4 (1973): 281–84. The same molecule is used by another species of whirligig, *Gyrinus natator*, from which it was independently isolated and characterized by H. Schildknecht, H. Neumaier, and B. Tauscher, "Gyrinal, die Pygidialrüsensubstanz der Taumelkäfer (Coleoptera: Carabidae)," *Justus Liebigs Annalen der Chemie* 756 (1972): 155–61.

I am grateful to Jerry Meinwald and Thomas Eisner for sharing their slides and knowledge with me, and to Fred McLafferty for providing an illustration of a mass spectrometer.

3. For more information on the uses of spectroscopy in structure determination, see Joesten, Johnston, Netterville, and Wood, *World of Chemistry*. See also "Signals from Within," program no. 10 in the Annenberg/Corporation for Public Broadcasting telecourse *The World of Chemistry*, available on videotape from Annenberg/CPB, P.O. Box 1922, Santa Barbara, Calif. 93116–1922.

4. For an evocative and readable description of the process of structure determination that stresses the elements of suspense and secrecy that make this part of chemistry an adventure, see the account by a master of the art in his autobiography: Carl Djerassi, *The Pill, Pygmy Chimps, and Degas' Horse* (New York: Basic Books, 1992), pp. 82–84.

4. Fighting Reductionism

1. For an introduction to the history of reductionism see E. Nagel, *The Structure of Science: Problems in the Logic of Scientific Explanation* (New York: Harcourt, Brace, and World, 1961). A cogent distinction between different kinds of reductionism is made by E. Mayr, *The Growth of Biological Thought* (Cambridge: Harvard University Press, 1982), pp. 59–64.

2. See Roald Hoffmann, "Nearly Circular Reasoning," *American Scientist* 76 (1988): 182–85.

3. See also Mary Jo Nye, *From Chemical Philosophy to Theoretical Chemistry: Dynamics of Matter and Dynamics of Disciplines* (Berkeley: University of California Press, 1993), chap. 10. For a contrasting (and, in my opinion, wrongheaded) view of chemistry as a "reduced science" in our century, see David Knight, *Ideas in Chemistry* (New Brunswick, N.J.: Rutgers University Press, 1992), chap. 12.

4. Let there be no doubt that I disagree strongly here with Steven Weinberg's spirited defense of reductionism, in his *Dreams of a Final Theory* (New York: Pantheon, 1992), esp. chap. 3.

5. The Fish, the Worm, and the Molecule

1. J. W. Cornforth, "The Trouble with Synthesis," *Australian Journal of Chemistry* 46 (1993): 157–70.

6. Telling Them Apart

1. Rosamond Smith, *Lives of the Twins,* p. 235.

7. Isomerism

1. Two quite readable introductions to chemistry are Joesten, Johnston, Netterville, and Wood, *World of Chemistry;* and Peter W. Atkins, *Molecules* (New York: Scientific American Library, 1987). See also Roald Hoffmann and Vivian Torrence, *Chemistry Imagined* (Washington, D.C.: Smithsonian Institution Press, 1993).

2. The number listed here does not include so-called geometrical and optical isomers (to be described below). These would show a still more fantastic increase—for $C_{20}H_{42}$ there are 3,395,964 isomers, for instance, if we count all types of isomerism.

3. W. C. Willett and A. Ascherio "Trans Fatty Acids: Are the Effects Only Marginal?" *American Journal of Public Health* 84 (1994): 722–24.

8. Are There Two Identical Molecules?

1. There *are* differences in the chemical behavior of isotopes. If there weren't any, we would not be able to separate them as easily as we can. So the boiling points of the diatomic molecules H_2 and T_2 differ by nearly five degrees C. Water enriched in deuterium is pretty easily prepared by electrolysis; the H_2/D_2 gas produced is depleted in deuterium. See N. N. Greenwood and A. Earnshaw, *Chemistry of the Elements* (Oxford: Pergamon, 1984), chap. 3, for further references.

For an amusing heavy-water episode see the travails of Chaplain Albert Tappman in Joseph Heller's sequel to *Catch-22: Closing Time* (New York: Simon & Schuster, 1994).

2. The abundances are from Robert C. West, ed., *The CRC Handbook of Chemistry and Physics,* 53rd ed. (Cleveland, Ohio: Chemical Rubber Company, 1972).

3. Henning Hopf, Braunschweig, private communication. Thanks to Mike Senko and Grisha Vajenine for discussions of isotopomer abundance.

9. Handshakes in the Dark

1. For a quite readable account of this remarkable story, see Jean Jacques, *The Molecule and Its Double,* trans. Lee Scanlon (New York: McGraw-Hill, 1993).

2. Heinrich Wölfflin, "Über das Rechts und Links im Bilde," in *Gedanken zur Kunstgeschichte* (Basel: Benno Schwabe, 1940), pp. 82–96. Raphael's *Sistine Madonna* is reproduced by permission of the Staatliche Kunstsammlungen Dresden. The painting is in the Gemäldegalerie Alte Meister, Dresden.

3. I am grateful to David Harpp not only for the illustrations of the carvone enantiomers but also for introducing his "World of Chemistry" course to Cornell.

4. W. F. McKean, C. J. L. Lock, H. E. Howard-Lock, "Chirality in Antirheumatic Drugs," *Lancet* 338, nos. 8782–83 (1991): 1565.

5. David Harpp has brought to my attention the remarkably detailed discussion of dextro- and levomethorphan in Patricia D. Cornwell's novel *Body of Evidence* (New York: Avon, 1991), pp. 236–47.

10. Molecular Mimicry

1. For an account of what we know of hemoglobin, see Lubert Stryer, *Biochemistry,* 3d ed. (New York: Freeman, 1988), chap. 7.

2. See ibid., pp. 195, 363, for more on alcohol dehydrogenase. Other alcohol dehydrogenase "inhibitors" are being tried therapeutically: George A. Porter, "The Treatment of Ethylene Glycol Poisoning Simplified," *New England Journal of Medicine* 319, no. 2 (July 14, 1988): 109–10.

A sweet-tasting (and also poisonous) molecule close in its chemistry to ethylene glycol is diethylene glycol, $HOCH_2CH_2OCH_2CH_2OH$. This was added to wine in a notorious Austrian wine adulteration scandal of the mid-1980s: William Drozdiak, "Bonn Seizes Austrian Wine," *Washington Post,* July 12, 1985, p. A30.

3. L. S. Goodman and A. Gilman, *The Pharmacological Basis of Therapeutics,* 8th ed. (New York: Pergamon, 1990).

4. Erich Posner, in the entry on Domagk in C. C. Gillespie, ed., *Dictionary of Scientific Biography* 4:153–56 (New York: Scribner's, 1970–).

5. Lewis Thomas, *The Youngest Science* (New York: Viking, 1983), p. 35.

6. For a critique of the natural tendency to fall into militaristic metaphor in describing drug design and immune system action, and a relevant and interesting sequel in Jacob's story, see Roald Hoffmann and Shira Leibowitz, "Molecular Mimicry, Rachel and Leah, the Israeli Male, and the Inescapable Metaphor in Science," *Michigan Quarterly Review* 30, no. 3 (1991): 382–97.

7. Posner, in the entry on Domagk in Gillespie, ed., *Dictionary of Scientific Biography* 4:153–56.

8. For more information on curare, as well as a wealth of fascinating material on molecular mimicry in action, see the marvelous book by John Mann, *Murder, Magic, and Medicine* (Oxford: Oxford University Press, 1992).

PART TWO
The Way It Is Told

12. And How It Came to Be That Way

1. See E. Garfield in *Essays of an Information Scientist* (Philadelphia: ISI [Institute of Scientific Information] Press, 1981), pp. 394–400, and references therein.

2. See S. Shapin, "Pump and Circumstance," *Social Studies of Science* 14 (1984): 487; P. Dear, "Totius-in-Verba—Rhetoric and Authority in the Early Royal Society," *Isis* 76 (1985): 145; and F. L. Holmes, "Scientific Writing and Scientific Discovery," *Isis* 78 (1987): 220–35.

3. For a discussion of the evolution of science writing, see B. Coleman, "Science Writing: Too Good to Be True?" *New York Times Book Review,* September 27, 1987, p. 1; see also R. Wallsgrove, "Selling Science in the Seventeenth Century," *New Scientist* (December 24–31, 1987): 55.

13. Beneath the Surface

1. The phrase is used after J. Ziman, *Reliable Knowledge* (Cambridge: Cambridge University Press, 1978). There is no better, nor more humanistic, description of what science is and should be than this small volume.

2. David Locke, *Science as Writing* (New Haven: Yale University Press, 1993).

3. J. Derrida, in his essay "Signature Event Context," in *Margins of Philosophy,* trans. A. Bass (Chicago: University of Chicago Press, 1982), 307–30; published originally as *Marges de la philosophie* (Paris: Editions de Minuit, 1972), pp. 365–93.

4. L. F. Fieser and M. Fieser, *Style Guide for Chemists* (New York: Reinhold, 1960).

5. P. B. Medawar, "Is the Scientific Paper Fraudulent?" *Saturday Review,* August 1, 1964, pp. 42–43, also argues that the standard format of the scientific article misrepresents the thought processes that go into discovery.

14. The Semiotics of Chemistry

1. For an introduction to modern literary theories see T. Eagleton, *Literary Theory* (Minneapolis: University of Minnesota Press, 1983).

2. C. F. von Weizsäcker, *Die Einheit der Nature* (Munich: DTV, 1974), p. 61.

3. Antoine-Laurent Lavoisier, *Elements of Chemistry,* trans. Robert Kerr (New York: Dover, 1965), p. xiii.

4. Lavoisier, *Elements of Chemistry,* p. xiv.

5. B. L. Whorf, "Languages and Logic," in J. B. Carroll, ed., *Language, Thought, and Reality* (Cambridge: MIT Press, 1956), p. 236.

6. Pierre Laszlo, *La parole des choses* (Paris: Hermann, 1993).

7. See also N. J. Turro, "Geometric and Topological Thinking in Organic Chemistry," *Angewandte Chemie* 98 (1986): 872, and *Angewandte Chemie (International Edition in English)* 25 (1986): 882, for a description of the geometrical and topological information processing that goes on in organic chemistry.

15. What DOES That Molecule Look Like?

1. Roland Barthes, *The Empire of Signs,* trans. Richard Howard (New York: Hill and Wang, 1982); from *L'empire des signes* (Geneva: Skira, 1980).

2. But not really. What we actually have, of course, is a human being guiding a tool, which in turn was programmed by other human beings—not to mention that it was built by still other humans and their tools. I am grateful to Dennis Underwood and Don Boyd for their help with illustrations 38 through 41.

3. For methodological discussions of how models are used in chemistry see C. J. Suckling, K. E. Suckling, and C. W. Suckling, *Chemistry Through Models* (Cambridge: Cambridge University Press, 1978); C. Trindle, "The Hierarchy of Models in Chemistry," *Croatica Chemica Acta* 57 (1984): 1231; J. Tomasi, "Models and Modeling in Theoretical Chemistry," *Journal of Molecular Structure (Theochem)* 48 (1988): 273–92. And for the range of meanings of "model," see

the amusing comment by N. Goodman, in *Languages of Art,* 2d ed. (Indianapolis: Hackett, 1976), p. 171.

4. That there are many ways to look at a molecule's structure is, of course, well known to the chemical community; I am not saying anything new here. See, for instance, G. Ourisson, *L'Actualité Chimique* (January–February 1986): 41, and the remarkable, innovative book by David S. Goodsell, *The Machinery of Life* (New York: Springer, 1993).

16. Representation and Reality

1. H. Mizes, S. Park, W. A. Harrison, "Multiple-Tip Interpretation of Anomalous Scanning-Tunneling-Microscopy Images of Layered Materials," *Physical Review B* 36 (1977): 4491; G. Binnig, H. Fuchs, Ch. Gerber, H. Rohrer, E. Stoll, and E. Tosatti, "Energy-Dependent State-Density Corrugation of a Graphite Surface as Seen by Scanning Tunneling Microscopy," *Europhysics Letters* 1 (1986): 32. See also the general discussion of what we see and don't see with STM, in Roald Hoffmann, "Now for the First Time, You Can See Atoms," *American Scientist* 81 (1993): 11–12.

2. The matter is actually not well understood, but Andrei Tchougreeff and I, and Myung-Hwan Whangbo, S. N. Magonov, and their coworkers have theories about it. Ergo, my saying that there are good reasons . . .

3. This photograph is reproduced, with permission, from J. Vouvé, J. Brunet, P. Vidal, J. Marsal, *Lascaux en Périgord Noir* (Périgeux: Pierre Fanlac, 1982), p. 31.

4. For an introduction to Hockney's neocubist perspective see David Hockney, *Cameraworks* (New York: Knopf, 1984).

17. Struggles

1. The cartoons by C. Heller are reproduced by permission from R. Hoffmann, "Under the Surface of the Chemical Article," *Angewandte Chemie* 100 (1988): 1653–63; *Angewandte Chemie (Int. Ed. Eng.)* 27 (1988): 1593–1602.

2. For an account of the industrial setting of Domagk's discovery, see John E. Lesch, "Chemistry and Biomedicine in an Industrial Setting: The Invention of the Sulfa Drugs," in Seymour H. Mauskopf, ed., *Chemical Sciences in the Modern World,* pp. 158–215 (Philadelphia: University of Pennsylvania Press, 1993).

Within a few months of the publication of Domagk's paper a French group found that the simpler sulfanilamide was as active as the more complicated Prontosil molecule. But sulfanilamide itself had no patent protection, for its synthesis and some antibacterial properties had been published earlier. It has been even speculated that IG Farbenindustrie knew this but delayed publication until a patentable alternative, Prontosil, could be made: James Le Fanu

("What Stopped the Magic Bullet?" *New Scientist* [July 18, 1985]). To be fair, I have not seen the evidence for this theory.

PART THREE
Making Molecules

19. Creation and Discovery

1. J. Davy, *Fragmentary Remains, Literary and Scientific, of Sir Humphry Davy* (London: Churchill, 1858), p. 14, as cited by David Knight in his wonderful biography of Davy, *Humphry Davy: Science and Power* (London: Blackwell, 1993). See also the sensitive review of this book by Oliver Sacks in the *New York Review of Books,* November 4, 1993, p. 50.

2. For a relevant debate on reductionism see S. Weinberg, "Newtonianism, Reductionism, and the Art of Congressional Testimony," *Nature* 330 (1987): 433, and the resulting exchange of letters between Weinberg and E. Mayr in *Nature* 331 (1988): 475.

3. J. Gleick, *Genius* (New York: Vintage, 1993), p. 437. I am grateful to Alan Lightman for bringing this quotation to my attention through his review of Gleick's book in the *New York Review of Books,* December 17, 1992.

4. Johann Wolfgang von Goethe, *Elective Affinities,* trans. R. J. Hollingdale (Harmondsworth: Penguin, 1971), p. 53. As the translator remarks, the English misses the German wordplay on "Scheidung" (divorce) and "Scheidekünstler" (the traditional name for an analytical chemist).

5. See in this context G. Stent, "Prematurity and Uniqueness in Scientific Discovery," *Scientific American* 227 (December 1972): 84–93, and G. Stent, "Meaning in Art and Science," *Engineering and Science* (California Institute of Technology, September 1985): 9–18.

6. For leading references to the beautiful chemistry discussed so cursorily here see A. R. Battersby and E. McDonald, "Origin of the Pigments of Life: The Type-III Problem in Porphyrin Biosynthesis," *Accounts of Chemical Research* 12 (1979): 14; A. R. Battersby, "Biosynthetic and Synthetic Studies on the Pigments of Life," *Pure and Applied Chemistry* 61 (1989): 337, and "How Nature Builds the Pigments of Life," *Pure and Applied Chemistry* 65 (1993): 1113–22; L. Milgrim, "The Assault on B_{12}," *New Scientist* (September 11, 1993): 39–44.

7. M. Berthelot, *Chimie organique fondée sur la synthèse* (Paris: Mallet-Bachelier, 1860), vol. 2. See also J.-P. Malrieu, "Du devoilement au design," *L'Actualité Chimique* 3 (1987): ix; A. F. Bochkov and V. A. Smit, *Organicheskii Sintez* (Moscow: Nauka, 1987).

8. Primo Levi, *The Monkey's Wrench* (New York: Simon & Schuster, 1986); published originally as *La Chiave a Stella* (Turin: Guilio Einaudi editore s.p.a., 1978).

9. David P. Billington, "In Defense of Engineers," *Wilson Quarterly* 10, no. 1 (1986): 89.

10. Baruch S. Blumberg has stressed the role of fantasy in model building in "The Making of a Medical Television Documentary," *American Medical Writers Association Journal* 4, no. 2 (1989): 19–25.

11. For a discussion of metaphor in science see various articles, especially that by R. R. Hoffman, "Some Implications of Metaphor for Philosophy and Psychology of Science," in R. Dirvan and W. Paprotte, eds., *The Ubiquity of Metaphor* (Amsterdam: John Benjamin, 1985).

12. Wallace Stevens, *The Palm at the End of the Mind: Selected Poems and a Play* (New York: Vintage, 1971), pp. 166–68.

13. Mary Reppy, private communication.

14. Eliseo Vivas, *Creation and Discovery* (Chicago: Henry Regnery, 1955), p. 137.

15. Ibid., p. xiii.

16. Richard Moore, "Poetry and Madness," *Chronicles* 58 (1991): 57.

17. Roald Hoffmann, "The Devil Teaches Thermodynamics," in *The Metamict State* (Orlando: University of Central Florida Press, 1987), p. 3.

20. In Praise of Synthesis

1. The special role of chemistry, its contrast to parts of physics and resemblance to art and engineering, is taken up in the article by J.-P. Malrieu, "Du devoilement au design." He coins the apt descriptor of *technopoïese* to characterize chemistry.

2. H. H. Claassen, H. Selig, J. G. Malm, "Xenon Tetrafluoride," *Journal of the American Chemical Society* 84 (1964): 3593.

3. N. D. Bartlett, "Xenon Hexafluoroplatinate (V) $Xe^+[PtF_6]^-$," *Proceedings of the Chemical Society* (1962): 218.

4. L. Malatesta, L. Naldini, G. Simonetta, and F. Cariati, "Triphenylphosphine-Gold(0)/Gold(I) Compounds," *Coordination Chemistry Reviews* 1 (1966): 255; V. G. Albano, P. L. Bellon, M. Manassero, and M. Sansoni, "Intermetallic Pattern in Metal-Atom Clusters," *Chemical Communications* (1970): 1210; F. Cariati and L. Naldini, "Trianionoeptakis(triarylphosphine)undecagold," *Inorganica Chimica Acta* 5 (1971): 172–74. The crucial crystal structure that revealed the Au_{11} cluster was for a thiocyanate derivative: M. McPartlin, R. Mason, and L. Malatesta, "Novel Cluster Complexes of Gold(0)–Gold(I)," *Chemical Communications* (1969): 334.

5. For some astute observations on the contrast of creative chemistry in industry and the universities see G. S. Hammond, "The Three Faces of Chemistry," *Chemtech* (1987): 140–43.

21. Cubane, and the Art of Making It

1. P. E. Eaton and T. W. Cole, Jr., "Cubane," *Journal of the American Chemical Society* 86 (1964): 3157–58. Illustration 21.1 is adapted with permission from this article. Copyright © 1964 American Chemical Society.

2. J. W. Cornforth, "The Trouble with Synthesis," *Australian Journal of Chemistry* 46 (1993): 159.

3. R. B. Woodward, "Synthesis," in A. R. Todd, ed., *Perspectives in Organic Chemistry* (New York: Interscience, 1956), p. 155.

4. E. J. Corey, "General Methods for the Construction of Complex Molecules," *Pure and Applied Chemistry* 14 (1967): 30.

5. For a remarkable, spirited analysis of synthesis see Cornforth, "The Trouble with Synthesis." See also R. Hoffmann, "How Should Chemists Think?" *Scientific American* 268 (1993): 66–73.

23. Natural/Unnatural

1. Igor Stravinsky, *The Poetics of Music* (Cambridge: Harvard University Press, 1956), p. 29.

2. See also Hubert Markl, "Die Natürlichkeit der Chemie," in Jürgen Mittelstrass and Günter Stock, eds., *Chemie und Geisteswissenschaften* (Berlin: Akademie Verlag, 1992), pp. 139–57.

3. The credit usually goes to Friedrich Wöhler, for urea. But a convincing article by Douglas McKie (brought to my attention by Loren Graham, whom I thank) shows that Wöhler's synthesis did not convince all that many people: D. McKie, "Wöhler's 'Synthetic' Urea and the Rejection of Vitalism: A Chemical Legend," *Nature* 153 (1944): 608–9.

25. Why We Prefer the Natural

1. P. I. Tchaikovsky, *Complete Works*, vol. 41, ed. A. Dmitriev (Moscow: Government Musical Publishing House, 1950), p. 198.

2. Jean-Paul Malrieu, letter to Roald Hoffmann, December 1, 1993.

3. See Edward O. Wilson, *Biophilia: The Human Bond with Other Species* (Cambridge: Harvard University Press, 1984); and Edward O. Wilson and Stephen R. Kellert, eds., *The Biophilia Hypothesis* (Washington, D.C.: Island Press/Shearwater Books, 1993), esp. the chapter by Stephen Kellert. Jared Diamond's contribution to the latter volume wisely tempers our natural attraction to the biophilia hypothesis.

4. Laura Wood, private communication.

5. From "Singling & Doubling Together," in A. R. Ammons, *Selected Poems*, expanded ed. (New York: Norton, 1986), pp. 114–15.

PART FOUR
When Something Is Wrong

27. Thalidomide

1. Phillip Knightley, Harold Evans, Elaine Potter, and Marjorie Wallace, *Suffer the Children: The Story of Thalidomide* (New York: Viking, 1979).

2. Sjöström and Nilsson, *Thalidomide*, pp. 124–25.

3. Sjöström and Nilsson, *Thalidomide*, 96–97.

4. Knightley et al., *Suffer the Children*, p. 47.

5. Ethel Roskies, *Abnormality and Normality: The Mothering of Thalidomide Children* (Ithaca, N.Y.: Cornell University Press, 1972), p. 2.

6. Francisco Goya's *Mother Showing Her Deformed Child to Two Women*, in *The Black Border Album* (1803–1812), no. 23; in Pierre Gassier, *Francisco Goya: The Complete Albums* (New York: Praeger, 1973), p. 182.

7. See Sjöström and Nilsson, *Thalidomide;* Knightley et al., *Suffer the Children;* and Harvey Teff and Colin Munro, *Thalidomide: The Legal Aftermath* (Westmead, U.K.: Saxon House, 1976). Thalidomide is used today in some countries in the treatment of leprosy (hanseniasis). Despite precautions, this has led to a new wave of fetal malformations in Brazil: "Talidomida é distribuída sem bula em BF," *O Estado de S. Paulo*, May 5, 1994, p. 3; "Descobertas mais 24 Vítimas da Talidomida," ibid, May 20, 1994, p. A13.

8. Sjöström and Nilsson, *Thalidomide*, p. 176. The name of the pharmaceutical firm is spelled in this book as "Chaz-Pfeizer."

9. W. Lenz in J. M. Robson, F. M. Sullivan, and R. L. Smith, eds., *Symposium on Embryopathic Activity of Drugs* (London: J. and A. Churchill, 1965).

10. See W. H. DeCamp, "The FDA Perspective on the Development of Stereoisomers," *Chirality* 1 (1989): 2–6, for a balanced review. An early study showed equal teratogenicity for both enantiomers: S. Fabro, R. L. Smith, R. T. Williams, "Toxicity and Teratogenicity of Optical Isomers of Thalidomide," *Nature* 215 (1967): 296. The contrary results I choose to emphasize are by G. Blaschke, H. P. Kraft, K. Fickentscher, F. Kohler, "Chromatographische Racemattrennung von Thalidomid und teratogene Wirkung der Enantiomere," *Arzneimittelforschung* 29 (1979): 1640–42. See also W. Winter and E. Frankus, "Thalidomide Enantiomers," *Lancet* 339, no. 8789 (1992): 365.

11. McKean, Lock, and Howard-Lock, "Chirality in Antirheumatic Drugs," pp. 1565–68.

12. M. H. Browne, "Mirror-Image Chemistry Yielding New Products," *New York Times*, August 3, 1991, p. C1; S. C. Stinson, "Chiral Drugs," *Chemical and Engineering News* 70, no. 39 (September 28, 1992): 46–79; W. A. Nugent, T. V. RajanBabu, M. J. Burk, "Beyond Nature's Chiral Pool: Enatioselective Catalysis in Industry," *Science* 259 (January 22, 1993): 479–83.

13. These statistics are from Technology Catalysts International Corp., as cited in S. C. Stinson, "Market, Environmental Pressures Spur Change in Fine Chemicals Industry," *Chemical and Engineering News* 72, no. 20 (May 16, 1994): 10–14.

14. Primo Levi, *The Periodic Table*, trans. Raymond Rosenthal (New York: Schocken, 1984), p. 60.

28. The Social Responsibility of Scientists

1. S. Makonkawkeyoon, R. N. R. Limson-Pobre, A. L. Moreira, V. Schauf, and G. Kaplan, "Thalidomide Inhibits the Replication of Human Immunodeficiency Virus Type 1," *Proceedings of the National Academy of Sciences (USA)* 90 (1993): 5974.

PART FIVE
How, Just Exactly, Does It Happen?

29. Mechanism

1. H. Okabe and J. R. McNesby, "Vacuum Ultraviolet Photolysis of Ethane: Molecular Detachment of Hydrogen," *Journal of Chemical Physics* 34 (1961): 668–69.

2. See, e.g., K. R. Popper, *Conjectures and Reflections* (New York: Basic Books, 1962), *Objective Knowledge: An Evolutionary Approach* (Oxford: Clarendon Press, 1972), and *The Logic of Scientific Discovery* (New York: Harper and Row, 1965).

3. For problems with falsification see Lewis Wolpert, *The Unnatural Nature of Science* (Cambridge: Harvard University Press, 1993), pp. 94–100.

30. The Salieri Syndrome

1. T. S. Eliot, *Murder in the Cathedral*, part 1 (New York: Harcourt Brace and World, 1963), p. 44.

2. A similar view has been expressed by Lewis Wolpert, "Science's Negative Public Image—A Puzzling and Dissatisfying Matter," *The Scientist* (June 14, 1993): 11. See also Wolpert, *The Unnatural Nature of Science*, p. 89.

3. Peter Shaffer, *Amadeus* (New York: Harper and Row, 1981).

4. A. S. Pushkin, *Mozart and Salieri*, trans. Antony Wood (London: Angel, 1982).

5. It is not easy to disentangle man from myth. Stanley Sadie's *The New Grove Mozart* (New York: Norton, 1983) and Wolfgang Hildesheimer's *Mozart*, trans. Marion Faber (New York: Farrar Straus Giroux, 1982) are two excellent

biographies. William Stafford, *The Mozart Myths* (Stanford, Calif.: Stanford University Press, 1991), says, "Mozart was not a perfect human being, if such a thing is possible. His letters prove he could be waspish, snobbish, uncharitable and a liar" (p. 140). The scatological element in Mozart's love letters to his cousin, Maria Anna Thekla Mozart, is great fun.

I am grateful to Neal Zaslaw for leading me to these references.

6. P. Feyerabend, *Against Method: Outline of an Anarchist Theory of Knowledge* (London: Verso, 1978).

31. Static/Dynamic

1. Some of the essential pieces of the theory were seen much earlier by Daniel Bernoulli in his 1738 treatise *Hydrodynamica sive de viribus et motibus fluidorum commentarii.* I was reminded of this by Edgar Heilbronner, a onetime Basler, like Bernoulli.

2. For most everything you ever wanted to know about garlic and onions, see Eric Block, "The Organosulfur Chemistry of the Genus *Allium*—Implications for the Organic Chemistry of Sulfur," *Angewandte Chemie* 104 (1992): 1158; *Angewandte Chemie (Int. Ed. Engl.)* 31 (1992): 1135–78. I am grateful to Dr. Block for several discussions.

To calculate the distance traveled between collisions and the average number of collisions, one needs an estimate of the effective size of the molecule— a *collision diameter.* This can sometimes be estimated from an experiment that measures the viscosity of a molecule (but this information is not available for garlic odor components). So we (Dr. William Shirley and I) optimized theoretically the geometry of diallyl disulfide with the CAChe modeling system, and from that result estimate an approximate diameter of 8 Å for that molecule.

3. Edgar Heilbronner, private communication.

4. R. C. Miller and P. Kusch, "Velocity Distributions in Potassium and Thallium Atomic Beams," *Physical Review* 99 (1955): 1314–21. The experiment was first carried out not with molecules but with potassium and thallium atoms.

32. Equilibrium and Perturbing It

1. For more on what makes chemical reactions go, see P. W. Atkins, *Atoms, Electrons, and Change* (New York: Scientific American Library, 1991).

2. The catalysts used today are an iron oxide with silica, alumina, and KOH.

PART SIX
A Life in Chemistry

33. Fritz Haber

1. The definitive biography of Haber has just been published in German: Dietrich Stoltzenberg, *Fritz Haber: Chemiker, Nobelpreisträger, Deutscher, Jude* (Weinheim: VCH, 1994). I am grateful to Peter Gölitz for making available to me a section of this biography prior to its publication. There is an earlier biography by Morris Goran, *The Story of Fritz Haber* (Norman: University of Oklahoma Press, 1967), and a novel on his life by Herman Heinz Wille, *Der Januskopf* (Berlin: Buch Club 65, 1970). There is also a chapter on Haber in Richard Willstätter's autobiography *From My Life* (New York: W. A. Benjamin, 1965). Haber's scientific work is beautifully reviewed in a memorial lecture by J. E. Coates, printed in the *Journal of the Chemical Society* (1939): 1645.

A most perceptive analysis of the life of this great chemist, ably set in his tumultuous times, is the chapter by his godson and outstanding scholar of European history, Fritz Stern, in *Dreams and Delusions* (New York: Knopf, 1987), pp. 51–76.

2. Karl Friedrich Bonhoeffer, in *Chemiker Zeitung* 58 (1934). Stern, *Dreams and Delusions*, p. 294, remarks, "It was an act of courage [in 1934] to publish an obituary about a Jewish chemist, an act characteristic of Bonhoeffer and of his entire family, which behaved so heroically and suffered so cruelly under the Nazis."

3. Stern, *Dreams and Delusions*, pp. 55–56.

4. Goran, *The Story of Fritz Haber*, p. 23.

5. For further information on Carl Bosch, see George B. Kauffman, "Two High-Pressure Nobelists," *Today's Chemist* 3, no. 4 (1990): 20–21.

6. L. F. Haber, *The Poisonous Cloud* (Oxford: Clarendon Press, 1986), p. 21.

7. Haber, *The Poisonous Cloud*, p. 27.

8. Ibid., p. 34.

9. Ibid., p. 244.

10. Wilfred Owen, "Dulce et Decorum Est," in Alexander W. Allison et al., eds., *The Norton Anthology of Poetry*, 3d ed. (New York: Norton, 1983), p. 1037.

11. Haber, *The Poisonous Cloud*, p. 242.

12. An interesting point raised by Mary Reppy is that chemists do not seem to have the deep feelings of guilt or responsibility about the development of "chemical" warfare that (some) physicists have had about the atomic bomb. Why? Reppy mentions three possibilities: (1) that chemical weapons were used earlier (and so we have had time to "forget"), (2) chemical weapons are nominally outlawed, and (3) the development of poison gases did not become "the boon/shaping moment" for chemistry the way the Manhattan Project was for physics (M. Reppy, private communication).

13. R. M. MacLeod, "Gold from the Sea: Archibald Liversidge, F.R.S., and the 'Chemical Prospectors': 1870–1970," *Ambix* 35, no. 2 (1988): 53–64.

14. MacLeod, "Gold from the Sea," p. 59. The Haber quote is from F. Haber, "Das Gold im Meerwasser," *Zeitschrift für Angewandte Chemie* 40 (1927): 303–14.

15. Stern, *Dreams and Delusions*, p. 73. The Einstein quotation is from a letter by Einstein to Haber dated May 19, 1933 (Einstein Papers, Boston).

16. Excerpt from Haber's letter of resignation, dated April 30, 1933, to the Nazi Minister of Science, Art, and Education (cited in full in Willstätter, *From My Life*, p. 289).

17. Stern, *Dreams and Delusions*, p. 74, citing a letter in the Max Planck archive.

PART SEVEN
That Certain Magic

34. Catalyst!

1. Richard Zare, Stanford University, private communication.

2. Goethe, *Elective Affinities*, pp. 33–34.

3. *New York Times Magazine*, November 28, 1993. The statement in the advertisement is nicely provocative to a chemist, for one of the things a catalyst does *not* do is to disturb the equilibrium. It changes things, to be sure, by speeding up the approach to equilibrium.

35. Three Ways

1. For a reasoned account of how much further we can go, and at what cost, see J. G. Calvert, J. B. Heywood, R. F. Sawyer, and J. H. Seinfeld, "Achieving Acceptable Air Quality: Some Reflections on Controlling Vehicle Emissions," *Science* 261 (July 2, 1993): 37–45.

2. My information here is drawn from a readable review by M. Shelef and G. W. Graham, "Why Rhodium in Automotive Three-Way Catalysts?" *Catalysis Reviews: Science and Engineering* 36, no. 3 (1994): 433–57. See also J. T. Kummer, "Use of Noble Metals in Automobile Exhaust Catalysts," *Journal of Physical Chemistry* 90 (1986): 4747–52.

3. Shelef and Graham, "Why Rhodium in Automotive Three-Way Catalysts?" p. 437.

4. These statistics come from *Platinum 1994* (London: Johnson-Matthey, 1994).

5. T. R. Ward, R. Hoffmann, and M. Shelef, "Coupling Nitrosyls as the First Step in the Reduction of NO on Metal Surfaces: The Special Role of Rhodium," *Surface Science* 289 (1993): 85–99.

6. M. Shelef, "Unanticipated Benefits of Automotive Emission Control: Reduction in Fatalities by Motor Vehicle Exhaust Gas," *Science of the Total Environment* 146–47 (1994): 93–101. ◢

7. D. Lester and R. V. Clark, "Toxicity of Car Exhaust and the Opportunity for Suicide: Comparison Between Britain and the United States," *Journal of Epidemiology and Community Health* 41 (1987): 117–20, and references therein.

36. Carboxypeptidase

1. For a description of the chemistry and structure of carboxypeptidase A, see Stryer, *Biochemistry*, pp. 215–20.

2. D. Kahne and W. C. Still, "Hydrolysis of a Peptide Bond in Neutral Water," *Journal of the American Chemical Society* 110 (1988): 7529–34.

3. W. N. Lipscomb, "Structure and Catalysis of Enzymes," *Annual Reviews of Biochemistry* 52 (1983): 17–34; D. W. Christianson and W. N. Lipscomb, "Carboxypeptidase A," *Accounts of Chemical Research* 22 (1989): 62–69.

4. D. E. Koshland, Jr., "Protein Shape and Biological Control," *Scientific American* 229, no. 4 (1973): 52–64.

5. Stryer, *Biochemistry*, p. 220 (emphasis in original).

6. Mircea Eliade, *The Forge and the Crucible*, trans. Stephen Corrin (New York: Harper and Row, 1962; 2d ed., Chicago: University of Chicago Press, 1978).

PART EIGHT
Value, Harm, and Democracy

37. Tyrian Purple, Woad, and Indigo

1. Much of my discussion in this chapter is drawn from E. Spanier, ed., *The Royal Purple and the Biblical Blue, Argaman and Tekhelet: The Study of Chief Rabbi Dr. Isaac Herzog on the Dye Industries in Ancient Israel and Recent Scientific Contributions* (Jerusalem: Keter Publishing, 1987). For a modern account of the ancient royal purple industry see P. E. McGovern and R. H. Michel, "Royal Purple Dye: The Chemical Reconstruction of the Ancient Mediterranean Industry," *Accounts of Chemical Research* 23 (1990): 152–58.

2. Gösta Sandberg, *Indigo Textiles* (Asheville, N.C.: Lark Books, 1989). I am grateful to Prof. Sandberg, who has a marvelous textile collection, for permission to reproduce illustrations 37.2 (top) and 37.3 from his book.

3. I am grateful to Prof. Dr. W. Rauh, Heidelberg, for providing me his beautiful photograph of a field of *Isatis tinctoria*, illustration 37.2, bottom.

4. R. M. Smith, J. J. Brophy, G. W. K. Cavill, and N. W. Davies, "Iridodials and Nepetalactone in Defensive Secretion of the Coconut Stick Insects, *Graeffea*

crouni," *Journal of Chemical Ecology* 5 (1979): 727; T. Eisner, "Catnip: Its Raison d'Etre," *Science* 146 (December 4, 1964): 1318–20; T. Eisner, D. F. Wiemer, L. W. Haynes, and J. Meinwald, "Lucibufagins: Defensive Steroids from the Fireflies *Photinus ignitus* and *P. marginellus* (Coleoptera: Lampyridae)," *Proceedings of the National Academy of Sciences (USA)* 75 (1978): 905–8; K. Nakanishi, T. Goto, S. Itô, S. Natori, and S. Nozoe, eds., *Natural Products Chemistry*, vol. 1 (Tokyo: Kodansha, 1974), pp. 469–75.

38. Chemistry and Industry

1. For a history of the origins of the synthetic dyestuffs industry see A. S. Travis, *The Rainbow Makers* (Bethlehem: Lehigh University Press, 1993).

2. M. L. Dertouzos, Richard K. Lester, R. M. Solow, and the MIT Commission on Industrial Productivity, eds., *Made in America: Regaining the Productive Edge* (Cambridge: MIT Press, 1989), p. 7. Illustration 38.3 is adapted from this reference, © The Massachusetts Institute of Technology.

39. Athens

1. For a discussion of Athenian democracy see J. M. Moore, *Aristotle and Xenophon on Democracy and Oligarchy* (Berkeley: University of California Press, 1986); Aristotle, *The Athenian Constitution*, trans. P. J. Rhodes (Harmondsworth: Penguin, 1984); and Mogens H. Hansen, "Was Athens a Democracy?" *Det Kongelike Danske Videnskapernes Selskab, Historisk filosofiske Meddelelser* 59 (1989): 2–47. I am indebted to Professor Lynne S. Abel for leading me to these sources."

41. Environmental Concerns

1. For contrasting views on the Alar controversy see Bradford H. Sewell, Robin M. Whyatt, Janet Hathaway, and Lawrie Mott, *Intolerable Risk: Pesticides in Our Children's Food* (New York: Natural Resources Defense Council, February 27, 1989); and Joseph D. Rosen, "Much Ado About Alar," *Issues in Science and Technology* (Fall 1990): 85–90. See also Eliot Marshall, "A is for Apple, Alar, and . . . Alarmist?" *Science* 254 (October 4, 1991): 20–21; E. M. Whelan, *Toxic Terror: The Truth Behind the Cancer Scares*, 2d ed. (Buffalo, N.Y.: Prometheus, 1993); and K. R. Foster, D. E. Bernstein, P. W. Huber, eds., *Phantom Risks: Scientific Inference and the Law* (Cambridge: MIT Press, 1993).

42. Science and Technology in Classical Democracy

1. Thucydides, *The Peloponnesian War,* trans. John H. Finley, Jr. (New York: Modern Library, 1942), p. 105.

2. See John F. Healy, *Mining and Metallurgy in the Greek and Roman World* (New York: Thames and Hudson, 1978), and references therein. I'm grateful to Peter Gaspar for introducing me to this valuable source of information.

3. I have read I. F. Stone's spirited and reasoned rationalization of the conviction of Socrates, his attempt to give "the Athenian side of the story, to mitigate the city's crime and thereby remove some of the stigma the trial left on democracy and on Athens" (I. F. Stone, *The Trial of Socrates* [New York: Little, Brown, 1988]). Stone is one of my heroes, and his story is a marvelous reconstruction of the Athens of the day. But I'm not convinced by it.

44. A Response to Worries About the Environment

1. Philip H. Abelson, "Toxic Terror: Phantom Risks," *Science* 261 (July 23, 1993): 407.

2. For discussions of risk assessment and perception see Paul Slovic, "Perception of Risk," *Science* 236 (April 17, 1987): 280–85; and Milton Russell and Michael Gruber, "Risk Assessment in Environmental Policy-Making," *Science* 236 (April 17, 1987): 286–90. See also Daniel Goleman, "Hidden Rules Often Distort Ideas of Risk," *New York Times,* February 1, 1994, p. C1.

3. Peter M. Sandman, "Risk Communication: Facing Public Outrage," *EPA Journal* (November 1987): 21–22. I have benefited from a correspondence on these issues with Peter Sandman.

4. Ibid.

5. Ann P. Kinzig and Robert H. Socolow, "Human Impacts on the Nitrogen Cycle," *Physics Today* 47 (November 1994): 24–31.

45. Chemistry, Education, and Democracy

1. Jeremy Bernstein, one of our great writers of science, makes some points on science education for the nonscientist that are similar to mine. He speaks of cultural deprivation, technological bewilderment, and technological necessity as our imperatives in *Cranks, Quarks, and the Cosmos* (New York: Basic Books, 1993).

PART NINE
The Adventures of a Diatomic

46. C_2 in All Its Guises

1. For a carefully reasoned and readable account of discipline building at the frontier of chemistry and physics, see Nye, *From Chemical Philosophy to Theoretical Chemistry.*

2. The C_2 potential energy curves are drawn after P. P. Fougere and R. K. Nesbet, "Electronic Structure of C_2," *Journal of Chemical Physics* 44 (1966): 285–98.

3. The distances in C_2 come from K. P. Huber and G. Herzberg, *Molecular Spectra and Molecular Structure*, vol. 4, *Constants of Diatomic Molecules* (Princeton: Van Nostrand Reinhold, 1979).

4. R. E. LaPointe, P. T. Wolczanski, and J. F. Mitchell, "Carbon Monoxide Cleavage by $(silox)_3Ta$ (silox = $t-Bu_3SiO^-$)," *Journal of the American Chemical Society* 108 (1986): 6382–84. This is not the only simple L_nMCCML_n complex. Illustration 46.4 left is adapted by permission from this reference. Copyright © 1986, American Chemical Society.

5. M. I. Bruce, M. R. Snow, E. R. T. Tiekink, and M. L. Williams, "The First Example of a . . . Acetylide Dianion," *Journal of the Chemical Society, Chemical Communications* (1986): 701–702. See also C. J. Adams, M. I. Bruce, B. W. Skelton, and A. H. White, "Construction of Unusual Metal Clusters Using Dicarbon (C_2) as a Collar," ibid. (1993): 446–50.

6. G. Longoni, A. Ceriotti, R. Della Pergola, M. Manassero, M. Perego, G. Piro, and M. Sansoni, "Iron, Cobalt, and Nickel Carbide-Carbonyl Clusters by CO Scission," *Proceedings of the Royal Society of London*, ser. A, 308 (1982): 47–57.

7. A. Simon and E. Warkentin, "$Gd_{12}C_6I_{17}$—A Compound with Condensed, C_2-Containing Clusters," *Zeitschrift für Anorganische und Allgemeine Chemie* 497 (1983): 79.

8. M. A. Barteau and R. J. Madix, "Acetylenic Complex Formation and Displacement via Acid-Base Reactions on Ag(110)," *Surface Science* 115 (1982): 355–81 (1982); P. A. Stevens, T. H. Upton, J. Stöhr, and R. J. Madix, "Chemisorption-Induced Changes in the X-Ray-Absorption Fine Structure of Adsorbed Species," *Physical Reviews Letters* 67 (1991): 1653–56.

9. According to the National Research Council's Survey of Earned Doctorates (1991), Washington, D.C.

PART TEN

The Dualities That Enliven

48. Missing

1. See Thomas S. Kuhn, *The Essential Tension* (Chicago: University of Chicago Press, 1977), chap. 9.

2. Ibid., p. 237.

3. Robert K. Merton, *On the Shoulders of Giants: A Shandean Postscript* (New York: Harcourt Brace and World, 1965).

4. E. Garfield, *Citation Indexing: Its Theory and Application in Science, Technology, and Humanities* (New York: Wiley, 1979).

5. Robert G. Bergman, "Values in Science," lecture at the University of Toledo, May 20, 1987 (and private communication). Published in revised form as "Irreproducibility in the Scientific Literature: How Often Do Scientists Tell the Truth and Nothing But the Truth?" *Perspectives* 8, no. 2 (1989): 2–3.

6. For a further discussion of the ramifications of this duality see Ian Hacking, *Representing and Intervening* (Cambridge: Cambridge University Press, 1983).

7. See Roald Hoffmann and Shira Leibowitz, "Pure/Impure," *New England Review* 16, no. 1 (Winter 1994): 41–64.

8. See Gerald Holton, *Thematic Origins of Scientific Thought*, rev. ed. (Cambridge: Harvard University Press, 1988), esp. the introduction; Holton, *The Advancement of Science, and Its Burdens* (Cambridge: Cambridge University Press, 1986), esp. chap. 1; Holton, "On the Role of Themata in Scientific Thought," *Science* 188 (April 25, 1975): 328–34; G. Holton, *The Scientific Imagination: Case Studies* (Cambridge: Cambridge University Press, 1978), chap. 4; Holton, "Analisi/sintesi," in *Enciclopedia*, vol. 1, *Abaco-Astronomia* (Turin: Einaudi, 1977), pp. 3–33. See also Robert K. Merton, "Thematic Analysis in Science: Notes on Holton's Concept," *Science* 188 (April 25, 1975): 335–38.

9. Oppositions and polarities have figured as an analytic methodology in the work of many. Some leading references include: Maura G. Flannery, "Biology Is Beautiful," *Perspectives in Biology and Medicine* 35, no. 3 (1992): 422–35; the discussion of simplicity and complexity by J. S. Fruton, *A Skeptical Biochemist* (Cambridge: Harvard University Press, 1992), chap. 3; Keith Tayler, *The Logic of Limits* (Cambridge, U.K.: Haslingfield Press, 1992).

Crystal Woodward, in a fascinating, as yet unpublished analysis of the art and creativity of R. B. Woodward, sees dualities such as planning/flexibility, prediction/the unexpected, theory/experiment, thought/the tangible, and goal-oriented synthesis/serendipity as figuring importantly in the work of this outstanding chemist. See Crystal Woodward, "Art and Elegance in the Synthesis of Organic Compounds: Robert Burns Woodward," in Doris B. Wallace and Howard E. Gruber, eds., *Creative People at Work: Twelve Case Studies* (New York:

Oxford University Press, 1989); Crystal Woodward, "Le rôle du plaisir esthétique ou l'art dans la chimie organique dans l'oeuvre de R. B. Woodward," *L'Actualité Chimique* (December 1993): 63–70.

49. An Attribute of the Devil

1. The Pauli quotation is from Holton, *The Scientific Imagination*, pp. 148–49.

2. For an introduction to Hegel, see F. C. Beiser, ed., *The Cambridge Companion to Hegel* (Cambridge: Cambridge University Press, 1993), esp. the chapter by Michael Forster, "Hegel's Dialectical Method," pp. 130–70.

3. There are "polypolar" epistemologies, one of the more intriguing of which is Mallarmé's, based on paradox. See R. G. Cohn, *Modes of Art*, Stanford French and Italian Studies, no. 1 (Saratoga, Calif.: Anma Libri, 1975), esp. chap. 1.

4. Emily R. Grosholz, "Nature and Culture in *The Souls of Black Folk* and *The Quest of the Silver Fleece*," forthcoming.

5. Robert L. Stevenson, *Dr. Jekyll and Mr. Hyde, the Merry Men and Other Tales* (London, J. M. Dent, 1925), p. 61. An article by David Jones reminded me of this episode.

6. Avner Treinin, "In Praise of Dualities," *Scopus* 40 (1990): 54–56.

7. Kathryn S. March, "Weaving, Writing, and Gender," *Man (N.S.)* 18 (1983): 729–44.

50. Chemistry Tense, Full of Life?

1. A fascinating perspective on alchemy is to be found in Eliade, *The Forge and the Crucible*.

2. R. J. Hollingdale, in the preface to his translation of Goethe's *Elective Affinities*, p. 14. See also Uwe Pörksen, *Deutsche Naturwissenschaftssprachen* (Tübingen: Narr, 1986), pp. 97–125.

3. On the psychological significance of alchemy, see C. G. Jung, *Psychology and Alchemy*, trans. R. F. C. Hull (London: Routledge, 1953). For an introduction to Jung's work, see Anthony Storr, *Jung* (New York: Routledge, 1991).

51. Cheiron

1. Robert Graves, *The Greek Myths* (Baltimore: Penguin, 1958), 151.g. Some say Cheiron was the descendant of Nephele and Ixion (a thorn in Zeus's side), ibid., p. 63. The mythological information in this chapter derives from this source.

2. Aeschylus, *Prometheus Bound,* trans. Edith Hamilton in *Three Greek Plays* (New York: Norton, 1975), p. 115.

3. Ibid., p. 141.

4. For a fascinating introduction to the substantial literature on the deep significance of the Centaur, and the interpretation of a Centaur image of Nietzsche linking science and art, see Richard Klein, "The *Mētis* of Centaurs," *Diacritics* (Summer 1986): 2–13.

Acknowledgments

I have a special bond with Brookhaven National Laboratory. A summer spent there while I was in college, building a low-level counting system for ^{11}C, racing on a bicycle from the Cosmotron to the Chemistry Department shacks while carrying a load of rapidly decaying atoms—that cannot be forgotten. I didn't become a radiochemist, but I learned much from Jim Cumming and Gerhart Friedlander. And the excitement of that summer kept me in chemistry, kept me from being seduced away by the humanities.

Thirty-three years later I gave the Pegram lectures at Brookhaven. It was a real pleasure to return. I thank Betsy Sutherland and the Pegram Lecture Committee for their invitation, and my colleagues and friends there for their hospitality. Ed Lugenbeel at Columbia University Press kept gently after me to bring the lectures into published form. He has been a great editor.

This book contains some chapters that have been published previously, a few in obscure venues. Several derive from a valuable collaboration with Pierre Laszlo, on representation in chemistry, and published in *Angewandte Chemie*. Another chapter is drawn from an art/science/literature collaboration with Vivian Torrence called *Chemistry Imagined*. Several pieces were first published in my "Marginalia" column in *American Scientist*. There I owe much to my editors—Michelle Press, Sandra Ackerman, and Brian Hayes. One essay, "Natural/Unnatural" (which now contains roughly the material found in chapters 22–25 in this volume), was extensively edited, to its benefit, by a poet/philosopher with great insight, Emily Grosholz. I am also most grateful to Roy Thomas for his editing of the entire book. Teresa Bonner made invaluable contributions to the artistic design of this book—it has been a pleasure to work with her.

The most careful reader of my manuscript, a person who has made many suggestions for improvement, has been my wife, Eva B. Hoffmann. She also aided me in the final stages of production of the book, and I am very grateful to her for her care and help. Perhaps Eva's most important contribution is that she has helped me recognize that the people who voice environmental concerns are not attacking chemistry or chemists. Our environment is something precious that we *all* need to think about deeply, rationally and emotionally.

Many illustrations in this book were drawn by Jane Jorgensen. Over the years her drawings have graced and added value to my work. The photographs in this book, unless otherwise credited, were done by Cornell University Photography. Patricia Giordano helped much in the typing of the manuscript. Crucial revisions were made in the course of a sabbatical stay in the Chemistry Department at New York University; I thank my colleagues there for their support. My research group helped me with some research of a different kind.

One person at Cornell, Mary Reppy, has read the entire book with extraordinary perception and care and made many important suggestions. So did Columbia University Press's readers—Dick Zare (on whose advice I added the three chapters on catalysis), Loren Graham, William Frucht, Laura Wood, Robert Shapiro, and Robert Merton. Thoughtful comments on the entire book were also made by Henning Hopf (at whose suggestion I wrote the eighth chapter), Pierre Laszlo, Jean-Paul Malrieu, Lionel Salem, Alain Sevin, and Brian Sutcliffe, and by Tadgh Begley, Paul Houston, William N. Lipscomb, Peter Sandman, and Ben Widom on individual chapters. Still others supplied essential detailed information or drawings; these I acknowledge in the endnotes.

Several people deserve special mention: Peter Gölitz, always supportive and feeding me much material; Bruce Ganem, a resource on this or that biological story; Lubert Stryer, whose *Biochemistry* text I've mined; Ehud Spanier, whose parents came from the same town in Galicia where I was born, and who introduced me to the biblical blue, *tek-helet;* Jerry Meinwald and Tom Eisner, colleagues who do consistently inspiring research; Mordecai Shelef, who got me interested in NO_x reduction; and Lynne S. Abel, for guiding me to the literature on Greek democracy.

This book is dedicated to my teachers at Columbia College. I got it into my head to finish Columbia in three years, but into those three years I crammed in an incredible wealth of courses. The world opened up to me, perhaps more so in the humanities than in chemistry. That it opened up so was to the credit of Columbia's core curriculum, the Contemporary Civilization and Humanities sequences, the introductory history of art and music courses. In subsequent Columbia courses, I encountered an absolutely extraordinary group of teachers, who brought the world of the intellect, of literature, art, and science to me. I remember them, and this book is for them.

INDEX